Fluorescence Lifetime Imaging Ophthalmoscopy

Martin Zinkernagel • Chantal Dysli

Editors

Fluorescence Lifetime Imaging Ophthalmoscopy

 Springer

Editors
Martin Zinkernagel
Department of Ophthalmology
and Department of Clinical Research
Inselspital Bern University Hospital
University of Bern
Bern
Switzerland

Chantal Dysli
Department of Ophthalmology
and Department of Clinical Research
Inselspital Bern University Hospital
University of Bern
Bern
Switzerland

ISBN 978-3-030-22877-4   ISBN 978-3-030-22878-1   (eBook)
https://doi.org/10.1007/978-3-030-22878-1

This Springer imprint is published by the registered company Springer Nature Switzerland AG
The registered company address is: Gewerbestrasse 11, 6330 Cham, Switzerland

Foreword

Imaging technologies for retinal diseases have made considerable progress in the last decades. One of the cornerstones of retinal imaging is based on autofluorescence. Fundus autofluorescence (FAF), including quantitative fundus autofluorescence, which measures the intensity of autofluorescence of endogenous fluorophores, is nowadays an established imaging modality in clinical ophthalmology. However, fluorescence cannot only be quantified by its intensity but also by its lifetime. Fluorescence lifetime imaging provides information on how long a fluorophore remains, on average, in its excited state before returning to its ground state. In contrast to intensity imaging, fluorescence lifetime imaging depends on the molecular environment of the fluorophore but not on its concentration and therefore is able to provide entirely new insights into the pathology of retinal and macular diseases. With the introduction of the fluorescence lifetime imaging ophthalmoscopy (FLIO), our understanding of fluorescence lifetimes of endogenous retinal fluorophores has expanded tremendously and new impetus has been generated for research of various retinal diseases.

This book is meant to provide a review of current knowledge of fluorescence lifetime imaging of the retina and an overview of fluorescence lifetime findings in common retinal diseases, starting from techniques and principles to clinical applications of FLIO. The book has been written for anyone interested in retinal imaging and understanding fluorescence lifetime imaging or in learning more about it.

We would like to express our sincere appreciation to Heidelberg Engineering who have further refined fluorescence lifetime imaging of the retina based on a confocal scanning ophthalmoscope and have supported the advancement of FLIO with relentless dedication. We would also like to thank the staff at Springer Publishing for their professional commitment and efficiency to publish this book and to promote this emerging imaging technique.

Bern, Switzerland
April 2019

Chantal Dysli
Martin Zinkernagel

Contents

Abstract

Introduction

Fluorescence lifetime imaging ophthalmoscopy (FLIO) allows noninvasive in vivo measurement of autofluorescence lifetimes of natural fluorophores of the retina upon laser excitation. Beyond autofluorescence intensity, autofluorescence lifetimes provide further information about the metabolic state of the retina. By now, various publications and reports are available on FLIO. These include studies in healthy subjects, various degenerative and hereditary retinal diseases as well as laboratory studies in mouse models.

This book shall summarize the current state of knowledge in the field of fluorescence lifetime imaging in ophthalmology, and highlight the most important basic and clinical features of this technique.

Methods

Fluorescence lifetime imaging was performed using a fluorescence lifetime imaging ophthalmoscope (Heidelberg Engineering, Heidelberg, Germany), which includes an excitation laser (470 nm), two detection channels (short wavelength: 498–560 nm and long wavelength: 560–720 nm), and an infrared camera to correct for eye movements. Thereby, within the image acquisition time of about 2 minutes, an autofluorescence intensity image and corresponding lifetime map, and an infrared image are gained. Subsequently, individual lifetime parameters can further be analyzed. Data was correlated with clinical information and other imaging modalities such as color fundus images, optical coherence tomography, and others as appropriate.

Results and Discussion

This book includes a basic section about the background, the technique, and the image analysis. Subsequently, results of in vitro measurements of individual fluorophores and compounds, and measurements in tissues, healthy eyes, and various retinal diseases are presented, interlinked, and discussed. At the end, possible future directions are outlined.

Conclusion

Fluorescence lifetime imaging ophthalmoscopy represents an interesting tool for early detection of retinal metabolic changes in various diseases. It can be used for diagnostic purposes as well as for follow-up examinations, and potentially provides a targeted tool for future interventional trials.

Abbreviations

A2E	N-retinylidene-N-retinyl-ethanolamine
AEL	Accessible exposure limit
AGE	Advanced glycation end products
AMD	Age-related macular degeneration
Anti-VEGF	Vascular endothelial growth factor inhibitors
BRB	Blood–retina barrier
CHM	Choroideremia
CHF	Complement factor H
CSCR	Central serous chorioretinopathy
D	Diopter
ERG	Electroretinography
ETDRS	Early Treatment of Diabetic Retinopathy Study
FA	Fluorescein angiography
FAD/FADH$_2$	Oxidized and reduced flavin adenine dinucleotide
FAF	Fundus autofluorescence
FLIM	Fluorescence lifetime imaging microscopy
FLIO	Fluorescence lifetime imaging ophthalmoscopy
FMN	Flavin mononucleotide
FRET	Förster resonance energy transfer
FWHM	Full width at half maximum
GA	Geographic atrophy
HEYEX	Heidelberg Eye Explorer
ICG	Indocyanine green angiography
IR	Infrared
IRF	Instrument response function
LSC	Long spectral channel (560–720 nm)
L	Lutein
MacTel	Macular telangiectasia type 2
MH	Macular hole
MNU	N-methyl-N-nitrosourea
MP	Macular pigment

MPV	Macular pigment volume
MZ	*Meso*-zeaxanthin
NAD+/NADH	Oxidized and reduced nicotinamide adenine dinucleotide
NaIO$_3$	Sodium iodate
NIR	Near infrared
NPDR	Nonproliferative diabetic retinopathy
OCT	Optical coherence tomography; SD-OCT: spectral domain
PED	Pigment epithelial detachment
qAF	Quantitative fundus autofluorescence
ROC	Receiver operating characteristic
ROI	Region of interest
ROS	Reactive oxygen species
RP	Retinitis pigmentosa
RPE	Retinal pigment epithelium
SSC	Short spectral channel (498–560 nm)
STED	Stimulated emission depletion
STGD	Stargardt disease
TCA	Tricarboxylic acid cycle
TCSPC	Time-correlated single photon counting
τ_m	Mean autofluorescence lifetime
UV	Ultraviolet
VEGF	Vascular endothelial growth factor
Z	Zeaxanthin

Chapter 1
Introduction

Martin Zinkernagel, Chantal Dysli, and Sebastian Wolf

FLIO is a noninvasive imaging modality based on fundus autofluorescence (FAF) intensity imaging. With each FLIO measurement, FAF intensity images are obtained simultaneously. In addition, the time between the excitation of the retinal fluorescence and the detection of fluorescence signal is recorded with FLIO, which is referred to as fluorescence or FAF lifetime or decay time. Imaging FAF lifetimes holds multiple advantages over only imaging the FAF intensity. FAF intensity images are dominated by the strong autofluorescence of lipofuscin. Other fluorophores with weaker FAF intensities may not be differentiated from stronger fluorophores with this imaging modality, although these weak fluorophores may play a crucial role in the pathophysiology of retinal diseases. In addition, changes in the distribution of such fluorophores may provide potential markers for early metabolic changes and therefore reveal additional information about retinal diseases.

The FLIO technique has originated from fluorescence lifetime imaging microscopy (FLIM) where the approach is used for detection of changes in the cellular environment such as the oxidation level, the pH value and protein binding stages [1]. This technique was originally adapted for the application in ophthalmology by Schweitzer et al. in 2002 [2]. Since then the technique has been further developed by Heidelberg Engineering and is now established for measurements in healthy eyes as well as in retinal diseases [3]. The current state of research shows that FLIO provides promising additional information compared to other imaging modalities and provides insights into early pathophysiologic changes.

M. Zinkernagel (✉) · C. Dysli · S. Wolf
Department of Ophthalmology and Department of Clinical Research, Inselspital,
Bern University Hospital, University of Bern, Bern, Switzerland
e-mail: martin.zinkernagel@insel.ch

© Springer Nature Switzerland AG 2019
M. Zinkernagel, C. Dysli (eds.), *Fluorescence Lifetime Imaging
Ophthalmoscopy*, https://doi.org/10.1007/978-3-030-22878-1_1

This book shall summarize the current state of knowledge in the field of fluorescence lifetime imaging in ophthalmology, and highlight the most important basic and clinical features of this technique. It includes a basic section about the background, the technique and the image analysis. Subsequently, results of in vitro measurements of individual fluorophores and compounds, measurements in tissues, mouse models, healthy eyes, and various retinal diseases are presented, interlinked and discussed. Possible future directions are outlined.

References

1. Becker W, Bergmann A, Biskup C. Multispectral fluorescence lifetime imaging by TCSPC. Microsc Res Tech. 2007;70:403–9.
2. Schweitzer D, et al. [Time-correlated measurement of autofluorescence. A method to detect metabolic changes in the fundus]. Ophthalmologe. 2002;99(10):774–9.
3. Dysli C, et al. Fluorescence lifetime imaging ophthalmoscopy. Prog Retin Eye Res. 2017;60:120–43.

Chapter 2
FLIO Technique and Principles

Martin Zinkernagel and Chantal Dysli

Fluorescence lifetime imaging is ideally suited for observing fluorescing molecules within the retina. Fluorophores have a multitude of spectroscopic properties including specific excitation maxima, individual emission spectra, and specific fluorescence lifetimes. These lifetimes can be influenced and modified by the local metabolic environment of the fluorophores. Thereby, FLIO cannot only be used to obtain information about the fluorophores concentration, but also about the fluorophores molecular environment with high sensitivity and signal specificity. A schematic illustration of the setting of fluorescence lifetime imaging ophthalmoscopy is provided in Fig. 2.1. When a molecule absorbs a photon, it is transformed into an excited state [1]. In order to return from this excited state to the ground state it releases energy by emitting photons of longer wavelengths. This results in a fluorescence emission where intensity decreases exponentially over time. The resulting fluorescence decay is a single exponential function for homogenous populations of fluorophores. However, for heterogenous populations of molecules, such as in the retina or choroid, the decay is a multiexponential function, because more than one fluorophore specimen is present. In the currently available FLIO device, the depth resolution is approximately 300 μm, and therefore not high enough to separate the fluorescence signal for individual retinal layers. As such, the incoming fluorescence signal derives from the entire neurosensory retina, the retinal pigment epithelium as well as from the choroid and possibly the sclera. Therefore, the obtained fluorescence lifetime signal in FLIO is a mixture from a myriad of different fluorophores [2, 3].

M. Zinkernagel (✉) · C. Dysli
Department of Ophthalmology and Department of Clinical Research, Inselspital,
Bern University Hospital, University of Bern, Bern, Switzerland
e-mail: martin.zinkernagel@insel.ch

© Springer Nature Switzerland AG 2019
M. Zinkernagel, C. Dysli (eds.), *Fluorescence Lifetime Imaging Ophthalmoscopy*, https://doi.org/10.1007/978-3-030-22878-1_2

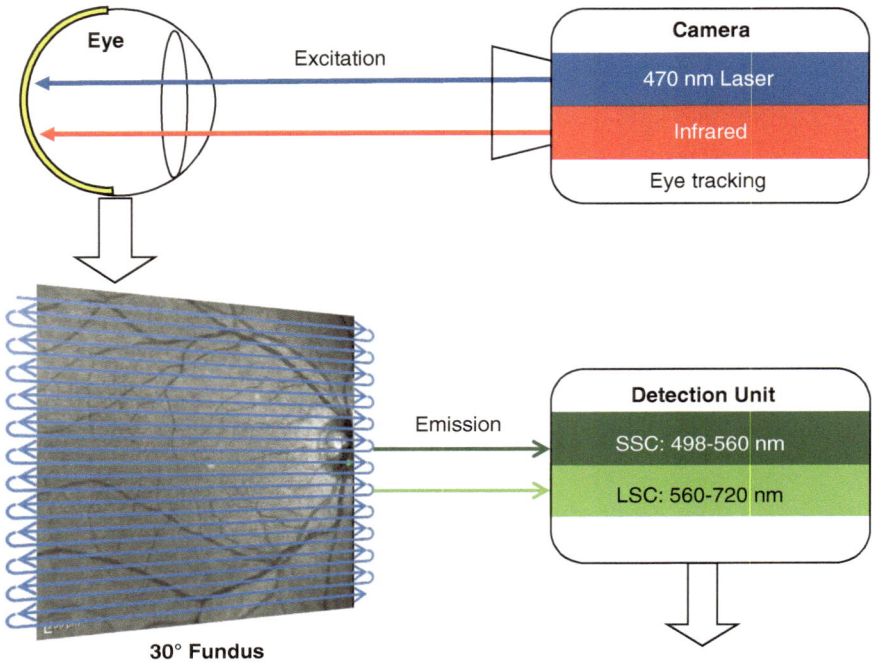

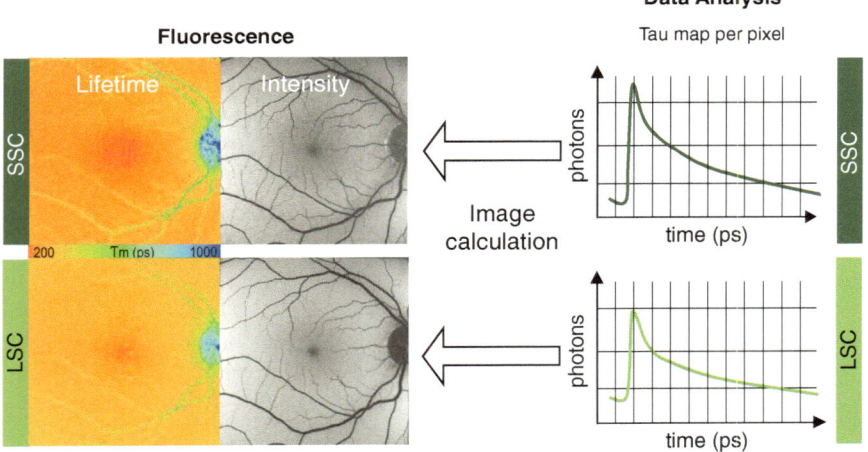

Fig. 2.1 Schematic illustration of the FLIO setup. For excitation of retinal autofluorescence, a 470 nm laser raster-scans the location of interest on the retina. An infrared camera is used for tracking of eye movements. Emitted fluorescence is registered in two distinct detection channels: a short (SSC, 498–560 nm) and a long (LSC, 560–720 nm) spectral channel. Corresponding decay curves are displayed for both channels. After calculation of the mean fluorescence lifetime (τ_m), a color coded lifetime map is displayed in parallel to the fluorescence intensity image for each channel

$$I(t) = I(0) * \sum_{i=1}^{n} \alpha_i e^{-\frac{t}{\tau_i}}$$

I: intensity
t: time
τ: lifetime
n: number of components
α: amplitude/weighting

Each emitted photon has to be registered in correlation to its arrival time. Furthermore, the exact location in the retina has to be recorded in order to obtain spatially resolved fluorescence lifetime data. In the FLIO device this is been achieved by employing a confocal laser scanning system for image registration. In order to obtain consistent time resolution, and resolution of multiexponential decay functions, a number of decay cycles with 12 ns interval are measured and sorted into multiple time channels. A theoretical decay function is plotted on the measured data from which decay components are then derived and fluorescence lifetimes finally calculated. In order to obtain an approximation for a higher order exponential decay, a higher number of photons is required. Furthermore, spatial resolution is restricted by the number of photons recorded. To reach a sufficient number of photons to achieve the statistical requirement to calculate an accurate decay approximation a higher binning factor of 1 or even 2 can be used. However, this is at the expense of spatial resolution of the FLIO image. If a sufficient number of photons has been recorded, a binning factor of 0 can be considered to calculate the lifetime for each single pixel in an image. Acquisition of at least 1000 photons is recommended in order to calculate an appropriate decay approximation [4].

There are several potential limitations for the currently used FLIO system. Because there may still be emission of photons present after the measurement cycle of 12 ns, or in other words, there may be incomplete decay of fluorescence, photons from the previous cycle may bleed into the next cycle and bias the lifetime measurements. Therefore, an incomplete decay model is used for fluorescence lifetime approximation. Other factors such as opacities of the optical media (cataract or corneal disease) may lead to scattering of incoming photons and influence lifetime measurements.

References

1. Becker W. Fluorescence lifetime imaging – techniques and applications. J Microsc. 2012;247(2):119–36.
2. Sauer L, et al. Review of clinical approaches in fluorescence lifetime imaging ophthalmoscopy. J Biomed Opt. 2018;23(9):1–20.
3. Dysli C, et al. Fluorescence lifetime imaging ophthalmoscopy. Prog Retin Eye Res. 2017;60:120–43.
4. Dysli C, et al. Quantitative analysis of fluorescence lifetime measurements of the macula using the fluorescence lifetime imaging ophthalmoscope in healthy subjects. Invest Ophthalmol Vis Sci. 2014;55(4):2106–13.

Chapter 3
FLIO Historical Background

Martin Hammer

Although fluorescence lifetime measurement is considered a relatively new technique in biomedical imaging (see Berezin and Achilefu for review [1]), it has been discovered in the nineteenth century already. In 1859 Edmond Bequerel developed the so called phosphoroscope with a time resolution of 10^{-4} s. In the 1920s, time resolution was improved to 10^{-8} s which enabled the first fluorescence lifetime measurements [2, 3]. However, only the advent of short pulse lasers and the introduction of time correlated single photon counting (TCSPC) [4, 5] made fluorescence lifetime measurement sufficiently sensitive for the detection of intrinsic fluorophores in living tissue. Fluorescence lifetime imaging microscopy (FLIM) evolved based on two different techniques: Firstly, a full field illumination, and the use of gated or streak cameras. This approach was pursued in frequency domain technique. Secondly, the time-domain approach in combination with confocal scanning laser microscopy. Specifically, two-photon excitation microscopy [6], using an inherently pulsed fluorescence excitation source, was used for FLIM investigations. Whereas FLIM of intrinsic fluorophores gives detailed information on cell metabolism [7] and may detect malignant changes [8–10], the development of genetically expressed fluorescent proteins resulted in further progress in structural as well as functional imaging [11]. Another milestone in fluorescence microscopy was the introduction of Förster resonance energy transfer (FRET) enabling the detection of interaction between labeled molecules [12]. Steady state FRET, however, relies on careful calibration. FLIM-FRET, however, allows to observe the molecular interaction directly from the quenching of the FRET donor [13, 14]. Recent developments in microscopy use fluorescence to overcome the diffraction limit of resolution with techniques such as stimulated emission depletion (STED) [15]. This can be combined with FLIM to obtain high resolution images providing a molecular signature of biological specimen [16, 17].

M. Hammer (✉)
University Hospital Jena, Department of Ophthalmology, Jena, Germany
e-mail: Martin.Hammer@med.uni-jena.de

© Springer Nature Switzerland AG 2019
M. Zinkernagel, C. Dysli (eds.), *Fluorescence Lifetime Imaging Ophthalmoscopy*, https://doi.org/10.1007/978-3-030-22878-1_3

Delori et al. were the first to measure fundus autofluorescence (FAF) spectra from single retinal locations [18]. First images of FAF were recorded by von Rückmann et al. in the 1990s [19–24]. As the age pigment lipofuscin, which accumulates in the retinal pigment epithelium (RPE) and is involved in the pathogenesis of age-related macular degeneration (AMD), was found to be a major retinal fluorophore, subsequent FAF studies addressed this disease. FAF was used to describe the progression of geographic atrophy of the RPE [25–29] and different patterns of FAF distribution were found [30–32]. This revealed the association of specific fluorescence patterns to sub-types of AMD as well as to its progression. However, observing the distribution of fluorescence intensities over the image did not give a clue on single fluorophores which might be of pathogenetic relevance. In order to distinguish fluorophores, Schweitzer et al. developed fluorescence lifetime ophthalmoscopy (FLIO), a method to measure fluorescence decay time which is specific for fluorophores as well as their embedding matrix [33–39]. These authors first applied lifetime imaging to the human retina *in vivo* in 2001 [35]. They fiber-coupled a mode-locked argon-ion laser into a scanning ophthalmoscope (cLSO, Carl Zeiss, Jena, Germany) and used TCSPC for fluorescence detection. However, the lack of an image registration algorithm limited the time available for the recording of an image without motion artifacts to few seconds. This resulted in the registration of some hundred photons per pixel only. Despite the resulting low signal to noise ratio, first fluorescence lifetime images were recorded [34]. An offline registration of recorded images was introduced in 2002 [36], and first clinical experiments in patients with age-related macular degeneration (AMD) were published in 2003 using a picosecond diode laser as light source [37]. Although the resolution was still low due to limited memory of the TCSPC electronics (64 × 64 pixels with a size of 80 × 80 μm^2), the images clearly revealed an prolongation of lifetimes in age-related macular degeneration [38]. Extensive *in vitro* and histological studies were performed to identify the fluorophores seen in fundus autofluorescence images, and to measure their emission spectra as well as fluorescence lifetimes [39, 40]. Considerable progress was made with the use of the Heidelberg Retina Angiograph scanner (Heidelberg Engineering, Heidelberg, Germany), enabling an online image registration [41]. An industrially designed prototype device, based on the Heidelberg Engineering Spectralis scanner, was first used by Dysli et al. in 2014 [42]. Recent clinical research with this device addressed age-related macular degeneration [43–46], diabetic retinopathy [47, 48], macular telangiectasia [49], macular holes [50], albinism [51], glaucoma [52], Stargard's disease [53, 54], retinal artery occlusion [55], central serous chorioretinopathy [56], choroideremia [57], retinitis pigmentosa [58, 59], Alzheimer's disease [60], and macular pigment density and distribution [51, 61].

References

1. Berezin MY, Achilefu S. Fluorescence lifetime measurements and biological imaging. Chem Rev. 2010;110(5):2641–84.
2. Gottling PF. Determination of the time between excitation and emissionfor certain fluorescent solids. Phys Rev. 1923;22:566–73.

3. Gaviola E. The dacay-time of dye stuff fluorescence. Ann Phys. 1926;81:681.
4. Leskovar B, et al. Photon-counting system for subnanosecond fluorescence lifetime measurements. Rev Sci Instrum. 1976;47(9):1113–21.
5. Lewis C, et al. Measurement of short-lived fluorescence decay using single photon-counting method. Rev Sci Instrum. 1973;44(2):107–14.
6. Denk W, Strickler JH, Webb WW. Two-photon laser scanning fluorescence microscopy. Science. 1990;248(4951):73–6.
7. Skala MC, et al. In vivo multiphoton microscopy of NADH and FAD redox states, fluorescence lifetimes, and cellular morphology in precancerous epithelia. Proc Natl Acad Sci U S A. 2007;104(49):19494–9.
8. Walsh AJ, et al. Optical metabolic imaging identifies glycolytic levels, subtypes, and early-treatment response in breast cancer. Cancer Res. 2013;73(20):6164–74.
9. Walsh AJ, et al. Quantitative optical imaging of primary tumor organoid metabolism predicts drug response in breast cancer. Cancer Res. 2014;74(18):5184–94.
10. Walsh AJ, et al. Temporal binning of time-correlated single photon counting data improves exponential decay fits and imaging speed. Biomed Opt Express. 2016;7(4):1385–99.
11. Winkler K, et al. Ultrafast dynamics in the excited state of green fluorescent protein (wt) studied by frequency-resolved femtosecond pump-probe spectroscopy. Phys Chem Chem Phys. 2002;4(6):1072–81.
12. Chen Y, Periasamy A. Characterization of two-photon excitation fluorescence lifetime imaging microscopy for protein localization. Microsc Res Tech. 2004;63(1):72–80.
13. Becker W, et al. Fluorescence lifetime imaging by time-correlated single-photon counting. Microsc Res Tech. 2004;63(1):58–66.
14. Duncan RR, et al. Multi-dimensional time-correlated single photon counting (TCSPC) fluorescence lifetime imaging microscopy (FLIM) to detect FRET in cells. J Microsc. 2004;215:1–12.
15. Hell SW, Wichmann J. Breaking the diffraction resolution limit by stimulated-emission – stimulated-emission-depletion fluorescence microscopy. Opt Lett. 1994;19(11):780–2.
16. Auksorius E, et al. Stimulated emission depletion microscopy with a supercontinuum source and fluorescence lifetime imaging. Opt Lett. 2008;33(2):113–5.
17. Buckers J, et al. Simultaneous multi-lifetime multi-color STED imaging for colocalization analyses. Opt Express. 2011;19(4):3130–43.
18. Delori FC. Spectrometer for noninvasive measurement of intrinsic fluorescence and reflectance of ocular fundus. Appl Opt. 1994;33(31):7439–52.
19. von Rückmann A, Fitzke FW, Bird AC. Distribution of fundus autofluorescence with a scanning laser ophthalmoscope. Br J Ophthalmol. 1995;79:407–12.
20. von Rückmann A, Fitzke FW, Bird AC. Clinical application of in vivo imaging of fundus autofluorescence. Investig Ophthalmol. 1995;36(4):238.
21. von Rückmann A, Fitzke FW, Bird AC. In vivo fundus autofluorescence in macular dystrophies. Arch Ophthalmol. 1997;115(5):609–15.
22. von Rückmann A, Fitzke FW, Bird AC. Fundus autofluorescence in age-related macular disease imaged with a laser scanning ophthalmoscope. Invest Ophthalmol Vis Sci. 1997;38(2):478–86.
23. von Rückmann A, Fitzke FW, Bird AC. Distribution of pigment epithelium autofluorescence in retinal disease state recorded in vivo and its change over time. Graefes Arch Clin Exp Ophthalmol. 1999;237(1):1–9.
24. von Rückmann A, et al. Abnormalities of fundus autofluorescence in central serous retinopathy. Am J Ophthalmol. 2002;133(6):780–6.
25. Schmitz-Valckenberg S, et al. Analysis of digital scanning laser ophthalmoscopy fundus autofluorescence images of geographic atrophy in advanced age-related macular degeneration. Graefes Arch Clin Exp Ophthalmol. 2002;240(2):73–8.
26. Schmitz-Valckenberg S, et al. Fundus autofluorescence and fundus perimetry in the junctional zone of geographic atrophy in patients with age-related macular degeneration. Invest Ophthalmol Vis Sci. 2004;45(12):4470–6.
27. Schmitz-Valckenberg S, et al. Correlation between the area of increased autofluorescence surrounding geographic atrophy and disease progression in patients with AMD. Invest Ophthalmol Vis Sci. 2006;47(6):2648–54.

28. Schmitz-Valckenberg S, et al. Optical coherence tomography and autofluorescence findings in areas with geographic atrophy due to age-related macular degeneration. Invest Ophthalmol Vis Sci. 2011;52(1):1–6.
29. Schmitz-Valckenberg S, et al. Semiautomated image processing method for identification and quantification of geographic atrophy in age-related macular degeneration. Invest Ophthalmol Vis Sci. 2011;52(10):7640–6.
30. Bindewald A, et al. Classification of fundus autofluorescence patterns in early age-related macular disease. Invest Ophthalmol Vis Sci. 2005;46(9):3309–14.
31. Bindewald A, et al. Classification of abnormal fundus autofluorescence patterns in the junctional zone of geographic atrophy in patients with age related macular degeneration. Br J Ophthalmol. 2005;89(7):874–8.
32. Einbock W, et al. Changes in fundus autofluorescence in patients with age-related maculopathy. Correlation to visual function: a prospective study. Graefes Arch Clin Exp Ophthalmol. 2005;243(4):300–5.
33. Schweitzer D, et al. Tau-mapping of the autofluorescence of the human ocular fundus. Proc SPIE. 2000;4164:79–89.
34. Schweitzer D, Kolb A, Hammer M. Autofluorescence lifetime measurements in images of the human ocular fundus. Proc SPIE. 2001;4432:29–39.
35. Schweitzer D, et al. Basic investigations for 2-dimensional time-resolved fluorescence measurements at the fundus. Int Ophthalmol. 2001;23:399–404.
36. Schweitzer D, et al. Zeitaufgelöste Messung der Autofluoreszenz – ein Werkzeug zur Erfassung von Stoffwechselvorgängen am Augenhintergrund. Opthalmologe. 2002;99(10):774–9.
37. Schweitzer D, et al. Evaluation of time-resolved autofluorescence images of the ocular fundus. In: Diagnostic optical spectroscopy in biomedicine II. 24–25 June 2003, Munich, Germany. 2003.
38. Schweitzer D, et al. In vivo measurement of time-resolved autofluorescence at the human fundus. J Biomed Opt. 2004;9(6):1214–22.
39. Schweitzer D, et al. Towards metabolic mapping of the human retina. Microsc Res Tech. 2007;70(5):410–9.
40. Schweitzer D, et al. Interpretation of measurements of dynamic fluorescence of the eye. Boston: SPIE; 2007.
41. Hammer M, et al. In-vivo and in-vitro investigations of retinal fluorophores in age – related macular degeneration by fluorescence lifetime imaging. In: SPIE Photonics West. 2009, SPIE.
42. Dysli C, et al. Quantitative analysis of fluorescence lifetime measurements of the macula using the fluorescence lifetime imaging ophthalmoscope in healthy subjects. Invest Ophthalmol Vis Sci. 2014;55(4):2106–13.
43. Dysli C, et al. Fluorescence lifetimes of drusen in age-related macular degeneration. Invest Ophthalmol Vis Sci. 2017;58(11):4856–62.
44. Dysli C, Wolf S, Zinkernagel MS. Autofluorescence lifetimes in geographic atrophy in patients with age-related macular degeneration. Invest Ophthalmol Vis Sci. 2016;57(6):2479–87.
45. Sauer L, et al. Monitoring foveal sparing in geographic atrophy with fluorescence lifetime imaging ophthalmoscopy – a novel approach. Acta Ophthalmol. 2018;96(3):257–66.
46. Sauer L, et al. Patterns of fundus autofluorescence lifetimes in eyes of individuals with nonexudative age-related macular degeneration. Invest Ophthalmol Vis Sci. 2018;59(4):AMD65–77.
47. Schmidt J, et al. Fundus autofluorescence lifetimes are increased in non-proliferative diabetic retinopathy. Acta Ophthalmol. 2017;95(1):33–40.
48. Schweitzer D, et al. Fluorescence lifetime imaging ophthalmoscopy in type 2 diabetic patients who have no signs of diabetic retinopathy. J Biomed Opt. 2015;20(6):61106.
49. Sauer L, et al. Fluorescence lifetime imaging ophthalmoscopy: a novel way to assess macular telangiectasia type 2. Ophthalmol Retina. 2018;2(6):587–98.
50. Sauer L, et al. Monitoring macular pigment changes in macular holes using fluorescence lifetime imaging ophthalmoscopy. Acta Ophthalmol. 2017;95(5):481–92.
51. Sauer L, et al. Fluorescence lifetime imaging ophthalmoscopy (FLIO) of macular pigment. Invest Ophthalmol Vis Sci. 2018;59(7):3094–103.

52. Ramm L, et al. Fluorescence lifetime imaging ophthalmoscopy in glaucoma. Graefes Arch Clin Exp Ophthalmol. 2014;252(12):2025–6.
53. Solberg Y, et al. Retinal flecks in Stargardt Disease reveal characteristic fluorescence lifetime transition over time. Retina. 2019;39(5):1. https://doi.org/10.1097/IAE.0000000000002519.
54. Dysli C, et al. Fluorescence lifetime imaging in Stargardt disease: potential marker for disease progression. Invest Ophthalmol Vis Sci. 2016;57(3):832–41.
55. Dysli C, Wolf S, Zinkernagel MS. Fluorescence lifetime imaging in retinal artery occlusion. Invest Ophthalmol Vis Sci. 2015;56(5):3329–36.
56. Dysli C, et al. Fundus autofluorescence lifetimes and central serous chorioretinopathy. Retina. 2017;37(11):2151–61.
57. Dysli C, et al. Autofluorescence lifetimes in patients with choroideremia identify photoreceptors in areas with retinal pigment epithelium atrophy. Invest Ophthalmol Vis Sci. 2016;57(15):6714–21.
58. Dysli C, et al. Fundus autofluorescence lifetime patterns in retinitis pigmentosa. Invest Ophthalmol Vis Sci. 2018;59(5):1769–78.
59. Andersen KM, et al. Characterization of retinitis pigmentosa using fluorescence lifetime imaging ophthalmoscopy (FLIO). Transl Vis Sci Technol. 2018;7(3):20.
60. Jentsch S, et al. Retinal fluorescence lifetime imaging ophthalmoscopy measures depend on the severity of Alzheimer's disease. Acta Ophthalmol. 2015;93(4):e241–7.
61. Sauer L, et al. Impact of macular pigment on fundus autofluorescence lifetimes. Invest Ophthalmol Vis Sci. 2015;56(8):4668–79.

Chapter 4
Fluorescence Lifetime Imaging Microscopy

Martin Zinkernagel

Fluorescence lifetime imaging microscopy (FLIM) is an imaging technique based on the analysis of exponential decay rates of fluorophores. In FLIM, the contrast relays on the lifetime of individual fluorophores rather than their emission spectra and/or intensity [1]. The first reports of FLIM date back to the late 1990s [2]. In FLIM, a very short pulsed laser is used to excite a sample. For fluorescence detection, a laser-scanning confocal microscope uses a pinhole effect, blocking all light from outside the focus. The emitted photons from the fluorophores pass back through an objective lens, and are then spectrally separated using a dichroic beam splitter. After the beam splitter, photons are quantified by highly sensitive detectors (photodiode or photomultiplier). FLIM is based on the differences in the excited state decay rates from fluorescent samples.

Generally, light absorbing molecules emit photons when returning from an excited state (S1 or subsequent states) to their ground state (S0). The transition from the excited state to the ground state is characterized by several parameters, such as, (a) the fluorescence spectrum, (b) the ratio of the total number of emitted photons to the number of absorbed photons, and, (c) the fluorescence lifetime. The lifetimes of most naturally occurring fluorophores typically range within the order of several nanoseconds (10^{-9} s). In order to measure these lifetimes, it is essential to use excitation pulses which are considerably shorter than the decay time of the fluorescence. Most FLIM systems are equipped with pulsed lasers which have a picosecond (10^{-12} s) or femtosecond (10^{-15} s) pulse duration. According to the "Stokes shift", the emitted fluorescence has a longer wavelength and less energy compared to the excitation light. Gabriel Stokes postulated in 1852 that there is a difference in energy between the maxima of the absorption and emission spectra. The fluorescence

M. Zinkernagel (✉)
Department of Ophthalmology and Department of Clinical Research, Inselspital, Bern University Hospital, University of Bern, Bern, Switzerland
e-mail: martin.zinkernagel@insel.ch

© Springer Nature Switzerland AG 2019
M. Zinkernagel, C. Dysli (eds.), *Fluorescence Lifetime Imaging Ophthalmoscopy*, https://doi.org/10.1007/978-3-030-22878-1_4

lifetime is a very useful parameter in imaging fluorescence, as it represents the change in fluorescence over time and is highly sensitive to the environment.

The measured mean fluorescence lifetimes, τ_m, are mapped spatially within the microscope image. The acquired nanosecond excited state lifetime is independent of the fluorophores concentration, sample thickness, or excitation intensity. As such, the signal is more robust than intensity-based methods. Fluorescence lifetimes depend on environmental parameters such as pH, ion concentration, and molecular binding or proximity to other energy acceptors. Therefore, FLIM allows for functional imaging. Super-resolution techniques of fluorescence imaging microscopy even allow for spatial resolution of single molecules [3].

Most conventional FLIM microscope systems operate in the visible wavelength range. Some systems operate in the near infrared (NIR) range (400–900 nm) and the latest developments use deep ultraviolet (UV, 240 nm) [4] or shortwave infrared (IR, 1700 nm) wavelengths [5].

There are several techniques to record the incoming photons. In Time-Correlated Single Photon Counting (TCSPC), FLIM intensity values are recorded in subsequent time channels [6, 7]. These time channels contain the number of recorded photons after the excitation pulse, which allows to acquire the decay curve for an individual point or pixel in the image. Fourier-domain FLIM does not record individual photon detection events, but rather records the fluorescence signal as a waveform [8].

Currently, there are several approaches for image analysis in TCSPC FLIM. The simplest way to analyze the data is by using a single exponential decay model. In this model the data is represented by a single decay time. In most cases, however, the decay profiles are more complex and have to be modelled by two or three exponential functions. These models contain several decay components and amplitude coefficients. The FLIM image is then presented as a color-coded image where decay times, or decay components, are represented by different colors.

FLIM can also be used to differentiate fractions of the same fluorophore in diverse states of interaction with its environment [9]. Several properties of FLIM can be exploited to gain additional molecular information or information on the fluorophores environment. Förster resonance energy transfer [10], or FRET, is based on energy transfer from the first molecule, or the donor molecule, to the second molecule, or the acceptor. FRET results in a quenching of the donor fluorescence and therefore decreases the decay time of the donor molecule. Because the energy transfer rate decreases with distance on a molecular scale, this technique has been very useful in molecular cell biology as FRET allows verifying whether molecules are physically linked on a nanometer scale [11, 12]. The use of FRET is currently the most frequent FLIM application.

Other applications include fluorescence quenching by oxygen, which is a quencher for a large number of fluorophores [12] or fluorophore quenching by ions such as Ca^{2+} and Cl^-, both of which are found in vast amounts in the neuronal system.

In autofluorescence FLIM, autofluorescence serves as an intrinsic, label free contrast. Because a mixture of intrinsic fluorophores can usually be found in a tissue sample, the decay profiles of tissue autofluorescence are multiexponential with lifetime components ranging from below 100 ps to several nanoseconds. In this case, multiple lifetimes may be observed as there are more than one fluorophore present in the sample or there are different chemical interactions with one fluorophore. Autofluorescence FLIM shares many similarities with FLIO. However, in the future, other FLIM applications such as identification of fluorescent dyes or nanoparticles may be combined with FLIO imaging and may then find clinical application.

References

1. Dysli C, Wolf S, Berezin MY, Sauer L, Hammer M, Zinkernagel MS. Fluorescence lifetime imaging ophthalmoscopy. Prog Retin Eye Res. 2017;60:120–43.
2. Bugiel I, Konig K, Wabnitz H. Investigation of cells by fluorescence laser scanning microscopy with subnanosecond time resolution. Lasers Life Sci. 1989;3:47–53.
3. Cox S, Jones GE. Imaging cells at the nanoscale. Int J Biochem Cell Biol. 2013;45(8):1669–78.
4. De Jong CJ, et al. Deep-UV fluorescence lifetime imaging microscopy. Photonics Res. 2015;3(5):283–8.
5. Becker W, Shcheslavsky V. TCSPC FLIM in the wavelength range from 800 nm to 1700 nm (Conference Presentation). 2016.
6. Becker W, et al. Fluorescence lifetime images and correlation spectra obtained by multidimensional time-correlated single photon counting. Microsc Res Tech. 2006;69(3):186–95.
7. Becker W, Su B, Bergmann A. Fast-acquisition multispectral FLIM by parallel TCSPC. In: SPIE BiOS. 2009, SPIE.
8. Gratton E, Jameson DM, Hall RD. Multifrequency phase and modulation fluorometry. Annu Rev Biophys Bioeng. 1984;13(1):105–24.
9. Becker W. Fluorescence lifetime imaging – techniques and applications. J Microsc. 2012;247(2):119–36.
10. Förster T. Energy migration and fluorescence. J Biomed Opt. 2012;17:011002.
11. Patterson GH, Piston DW, Barisas BG. Forster distances between green fluorescent protein pairs. Anal Biochem. 2000;284(2):438–40.
12. Lakowicz JR, et al. Fluorescence lifetime imaging of free and protein-bound NADH. Proc Natl Acad Sci U S A. 1992;89(4):1271–5.

Chapter 5
The Fluorescence Lifetime Imaging Ophthalmoscope

Chantal Dysli, Martin Zinkernagel, and Sebastian Wolf

Since 2010, prototypical fluorescence lifetime imaging ophthalmoscopes are used for research purposes, which were developed by Heidelberg Engineering (Heidelberg, Germany). These devices are called FLIO (fluorescence lifetime imaging ophthalmoscope) and, by now, they are used for experimental as well as for clinical studies [1]. The current FLIO setup is refined, based on the first experimental FLIO device, which was built by Schweitzer et al. in 2002. For this first application, the FLIM technology (see Chap. 4) was adapted for the use in human eyes [2–4]. Further details about the historical background of FLIO are presented in Chap. 3.

The currently used FLIO device (Fig. 5.1) is constructed upon a Heidelberg Retina Angiograph cSLO (HRA2, Heidelberg Engineering) [5]. A schematic illustration of the technical setup is provided in Fig. 5.2. A 470 nm blue laser diode (Becker & Hickl) with picosecond pulsed illumination excitation is used for fluorescence excitation. The laser beam is designed to be confocal to excite retinal fluorophores and thereby minimize the fluorescence of other ocular components outside from the confocal plane. During image acquisition, the laser beam raster scans the fundus with a frame rate of 9 Hz and a repetition rate of 80 MHz. The laser beam emits pulses of 89 ps full width at half maximum (FWHM). The FLIO device is in accordance with the laser safety guidelines of ANSI (ANSI, L.I. o. A., 2007. ANSI Z136.1d2007 American National Standard for Safe Use of Lasers) and International Electrotechnical Commission (IEC, 2014. IEC International Electrotechnical Commission. Standard 60825e1:2007 (Edition 2, ISBN 2-8318-9085-3), third ed. ISBN 978-2-8322-1499-2). The nominal average laser power is $P0 = 200$ μW. Therefore the energy (E) per pulse would be 2.5 pJ, which results in a single pulse peak power P of 2.5×10^{-12} J/89×10^{-12} s = 28 mW. The size of the laser

C. Dysli (✉) · M. Zinkernagel · S. Wolf
Department of Ophthalmology and Department of Clinical Research, Inselspital,
Bern University Hospital, University of Bern, Bern, Switzerland
e-mail: chantal.dysli@insel.ch

© Springer Nature Switzerland AG 2019
M. Zinkernagel, C. Dysli (eds.), *Fluorescence Lifetime Imaging Ophthalmoscopy*, https://doi.org/10.1007/978-3-030-22878-1_5

Fig. 5.1 Photography of a
fluorescence lifetime
imaging ophthalmoscope.
A camera with a 30° lens is
used. Behind, the power
supply and detection unit is
visible below the screen

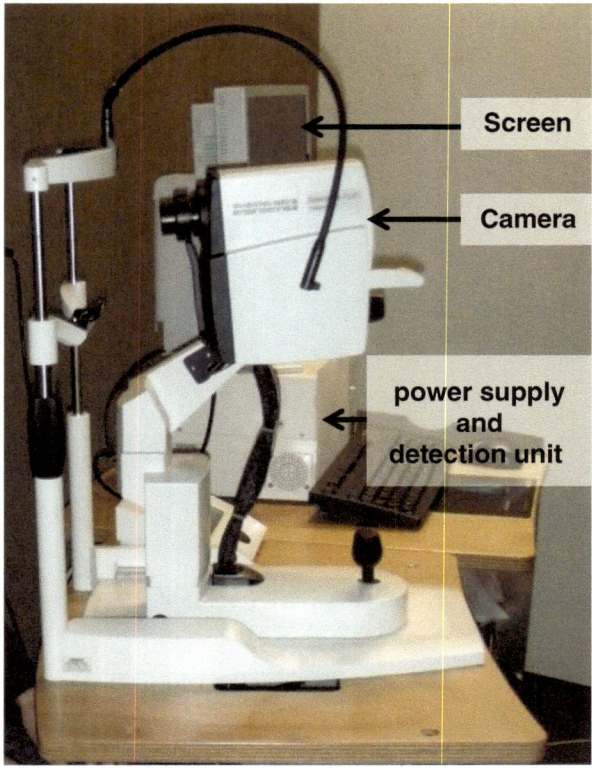

spot is approximately 75 μm² on the retina, which results in an intensity of 28 mW/75 μm²,
respectively 37 kW/cm². Due to the fast scanning procedure, only a few single laser
pulses (eight pulses with total energy of 20 pJ) sum up at one spot until proceeding to
the adjacent pixel, as the pixel clock of 10 MHz corresponds to 100 ns. Thereby, damage
of the retina applied with a continuous wave laser on a stationary retina spot can be
avoided. Additionally, the duty cycle for the extreme high-laser power density is very
low, that is, 89 ps/12.5 ns = 7 ∗ 10⁻³. The mean power, which is critical for thermal damage, is more than two orders of magnitude lower, (196 μW/75 μm² = 261 W/cm²).
According to the IEC 60825-1:2007 standard, the accessible exposure limit (AEL) for
class 1 laser products is deduced for a single 89-ps laser pulse at 470 nm as follows:

Accessible single pulse exposure limit:

$$AEL_{single} = 1.0 \times t^{0.75} \ \text{J} = 1.0 \times 8.9^{0.75} \times 10^{(-11 \times 0.75)} \ \text{J}$$
$$= 2.9 \times 10^{-8} \ \text{J}.$$

Accessible single pulse exposure for FLIO:

$$AE_{FLIOsingle} = 2.5 \times 10^{-12} \ \text{J}$$

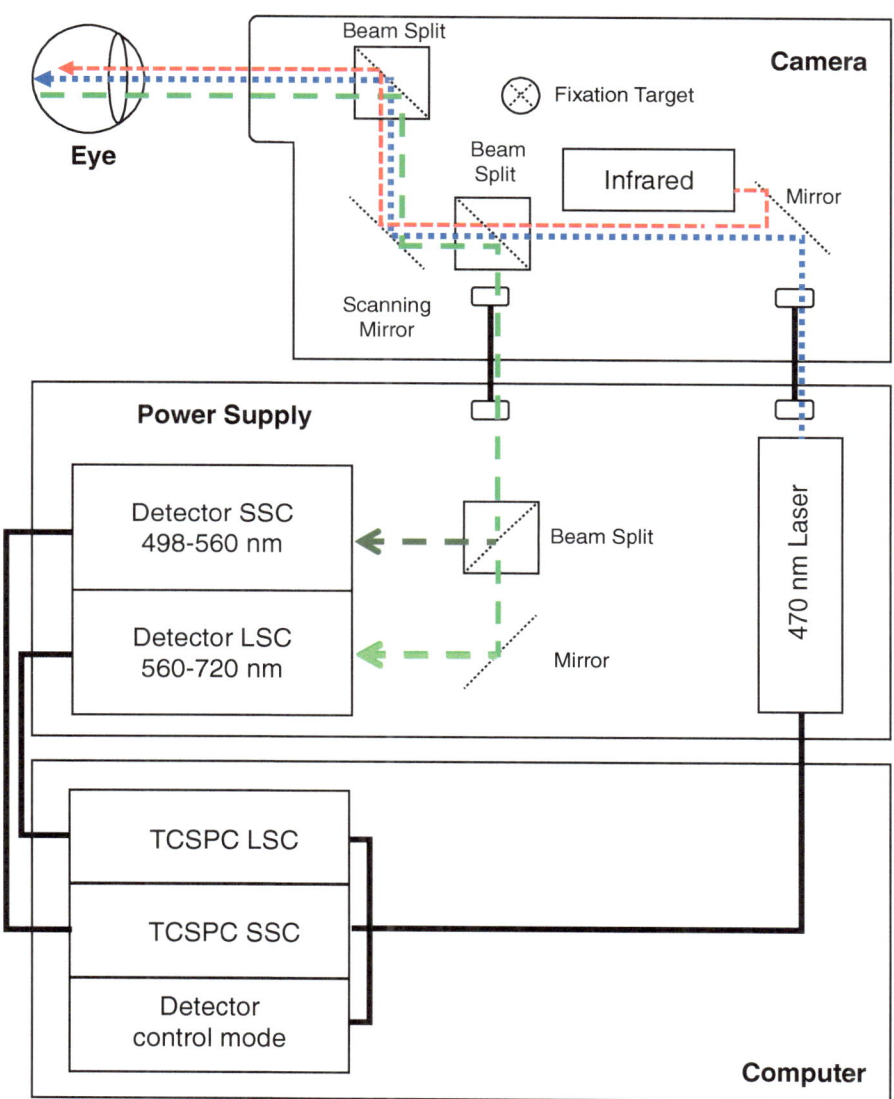

Fig. 5.2 Technical details of the fluorescence lifetime imaging ophthalmoscope. Details are described in the text. Retinal autofluorescence is excited with a 470 nm laser, and fluorescence emission is detected in two distinct detection channels for short (SSC) and long (LSC) emission wavelengths via time correlated single photon counting modules (TCSPC)

Therefore, the FLIO fulfills, with respect to single pulse exposure, the laser class 1 criteria by far. The pulse energy is almost a factor 104 smaller than the AEL for single pulses, defined by the standard.

The maximum permissible radiant power through the pupil for a pulsed exposure MPΦB, PLS for FLIO is approximately 2.8 mW. Therefore, FLIO is using an average laser power well below the limits recommended by the ANSI.

The standard setting of FLIO covers a field of 30° on the posterior pole. This refers to an area of approximately 9 × 9 mm at the human fundus. Normally, a central internal central fixation target is used resulting in an image section centered at the fovea. However, it is possible to adjust the position of the internal as well as an external fixation target in order to adapt the position of the FLIO image to the region of interest, e.g. to image the optic nerve as well as other pathologies that may not be found directly inside the macula.

Simultaneously to the fluorescence excitation, single emitted fluorescence photons are detected by two highly sensitive hybrid photon-counting detectors (HPM-100-40; Becker & Hickl GmbH, Berlin, Germany). The detectors rely on the principle of time-correlated single photon counting (TCSPC) [6, 7]. Fluorescence photons are recorded in two separate wavelengths channels. The short spectral channel (SSC) detects fluorescence photons with wavelengths between 498 and 560 nm, whereas the long spectral channel (LSC) records fluorescence photons with wavelengths between 560 and 720 nm. The channels were chosen by Heidelberg Engineering according to studies which assumed the LSC to be predominantly influenced by the emission of lipofuscin [3]. The SSC is believed to detect photons within a range where several retinal fluorophores emit fluorescence, including flavin adenine dinucleotide (FAD), advanced glycation end-products, collagen/elastin and macular pigment.

The 30° field imaging results in a 256 × 256 pixel detection frame. For each pixel, fluorescence excitation results in a repetitive stimulation every 12 ns, and detected photons are displayed in a logarithmic decay histogram (number versus timespan). The detected photons are recorded into 1024 time channels for each pixel, resulting in a photon arrival histogram representing the fluorescence decay. An inbuilt infrared camera is used for eye tracking and ensures that every single photon is detected at the correct special location within the detection frame. According to a standardized protocol, at least 1000 photons are usually recorded as a minimal signal threshold in the SSC at the foveal center.

The instrument response function (IRF) may slightly vary between individual devices. The practical minimal time resolution for the FLIO technique depends on the individual IRF and is as low as approximately 30 ps.

Image acquisition is operated over a modified Heidelberg Eye Explorer (HEYEX) software, and data is stored within the same platform. Data analysis and further lifetime calculation is done in an integrated version of SPCImage (Becker & Hickl) software. Further details can be found in Chap. 6.

References

1. Dysli C, et al. Fluorescence lifetime imaging ophthalmoscopy. Prog Retin Eye Res. 2017;60:120–43.
2. Schweitzer D, et al. [Time-correlated measurement of autofluorescence. A method to detect metabolic changes in the fundus]. Ophthalmologe. 2002;99(10):774–779.

3. Schweitzer D, et al. Basic investigations for 2-dimensional time-resolved fluorescence measurements at the fundus. Int Ophthalmol. 2001;23(4–6):399–404.
4. Schweitzer D, Kolb A, Hammer M. Autofluorescence lifetime measurements in images of the human ocular fundus. In: Proc SPIE, vol. 4432; 2001. p. 29–39.
5. Dysli C, et al. Quantitative analysis of fluorescence lifetime measurements of the macula using the fluorescence lifetime imaging ophthalmoscope in healthy subjects. Invest Ophthalmol Vis Sci. 2014;55(4):2106–13.
6. Becker W. The bh TCSPC handbook. 6th ed. Berlin: Becker & Hickl GmbH; 2014.
7. Becker W, et al. Fluorescence lifetime imaging by time-correlated single-photon counting. Microsc Res Tech. 2004;63(1):58–66.

Chapter 6
Image Acquisition and Analysis

Chantal Dysli, Lydia Sauer, and Matthias Klemm

To ensure optimal image quality, a standardized protocol for image acquisition and analysis is essential. Image acquisition is equivalent to other Heidelberg imaging modalities, especially FAF intensity imaging. However, there are a few important points to consider for acquisition of FLIO images. Generally, FLIO images are taken with maximally dilated pupils, [1, 2] which improves the detected fluorescence photon count rate, reduces the image acquisition time, and minimize the possible influence of lens artifacts. To reduce light scattering and to improve image quality, complete darkening of the room is recommended while obtaining the measurement. FLIO imaging must be performed before use of fluorescent dye, both topically for Goldmann applanation tonometry and systemically for fluorescein angiography (FA) or indocyanine green angiography (ICG).

Before image acquisition, subjects are placed in a comfortable position, using both, the chin and the forehead rest to ensure stable positioning. Proper opening of the eyes is important to reduce shadowing from the eyelids and eyelashes. Occasional blinking is allowed during the measurement and ensures a stable tear film and avoids corneal dryness, which would result in reduced image quality. For gaze stabilization, an internal or external fixation mark is used. Image acquisition in patients with nystagmus may still be possible due to the inbuilt infrared camera eye tracking system, however, FLIO image quality may be reduced in such cases.

C. Dysli (✉)
Department of Ophthalmology and Department of Clinical Research, Inselspital, Bern University Hospital, University of Bern, Bern, Switzerland
e-mail: chantal.dysli@insel.ch

L. Sauer
University of Utah, John A. Moran Eye Center, Salt Lake City, UT, USA

M. Klemm
Institute of Biomedical Engineering and Informatics, Technische Universität Ilmenau, Ilmenau, Germany

The illumination of the fundus should be homogeneously, and the camera focus should be adjusted for small to medium vessels. In general, a minimal photon count of 1000 photons per pixel in areas of low autofluorescence is collected in order to assure an adequate fitting procedure. This leads to an image acquisition time of approximately 90–120 seconds per eye. As macular pigment absorbs excitation as well as fluorescent light, subjects with high amounts of retinal carotenoids (e.g. on supplementation) may require a slightly longer acquisition time (120–150 seconds per eye). Patients with significant cataract, on the other hand, may require higher photon counts (1250–1500) to ensure good image quality. Lens fluorescence increases with cataract formation, leading to detection of relatively more fluorescence from the lens and less photons from the retina. A detailed image acquisition protocol is provided in the supplementary material.

A recent study investigated possible bleaching effects on the FLIO measurement [3]. No significant changes in FAF lifetimes were found for the SSC, and all lifetime changes in the LSC were <10% in magnitude. Interestingly, the acquisition time was shortened in the bleached state by approximately 20% and the fluorescence intensity was increased. Therefore, the authors concluded that bleaching does not present as a relevant confounder in the FLIO application. Furthermore, the measurement of recovered state was comparable to the baseline measurement. The amount of bleaching in this study exceeds by far any other imaging procedure performed in clinical routine. It is therefore unlikely that preceding retinal imaging significantly impacts the FLIO measurements in clinical routine. Nevertheless, FLIO imaging needs to be performed before any application of fluorescent dye as described before.

For FAF lifetime calculation and image analysis, FLIO raw data, composed of detected photons over time per pixel for both spectral channels, is imported in SPCImage software (Becker & Hickl). The original fluorescence lifetime data is used for approximation of an exponential decay curve in order to calculate a mean fluorescence lifetime value (τ_m). This approximation is also referred to as fitting procedure, since multiple exponential components are fitted to describe the fluorescence decay. Additionally, a device-specific instrument response function (IRF) is included to account for the device-specific beam laser path. For fitting of the fluorescence decay curve, multiple adaptations can be performed depending on the raw data set and the scientific question to be addressed. Generally, the following settings are used in previous and ongoing studies. However, fitting procedures may change if software and hardware technology improve.

1. To increase the number of photons per pixel and, thus, to improve the accuracy of the exponential fitting procedure, a binning of neighboring pixels is applied. An advantage of a higher binning factor is a potentially more accurate fit, however, the disadvantage lies within the loss of detail and spatial resolution. Therefore, depending on the scientific question, different settings may be used. Whereas studies taking an average lifetime value over large areas may be best conducted with a binning factor of 2 or higher, other studies investigating detailed and small changes in pathology may require not to use a binning factor at all.

Currently, in most clinical studies, a binning factor of 1 appears to be useful, which results in a moving average of a pixel-frame of 3×3 pixels. Thus, outlier values of individual pixels are minimized, and the lifetime values are smoothened without losing too much special resolution.

2. Within the fitting approach, the fluorescence decay curve can be described using one, two or three decay components. Fitting with one single decay is only applicable in pure substances where just one lifetime value is expected. As the retina yields contributions of multiple fluorophores, a multi-exponential decay model is preferred in this setting. Therefore, a two- or three-exponential decay model is commonly applied. The approximation can be described with the following equation:

$I(t)$ is the fluorescence intensity at time t and I_0 the initial, maximal intensity, α is the amplitude of each lifetime (τ) component, and $\otimes$ indicates the convolution integral with the IRF.

$$\frac{I(t)}{I_0} = IRF \otimes \sum_i \alpha_i \times e^{-\frac{t}{\tau_i}}$$

The goodness of fit can be quantified by the chi^2 value, whereby values towards 1.0 represent a better decay fit than higher values. As an additional second control, the SPCImage offers a residual plot which, similar to chi^2, also represents the goodness of fit.

The two exponential fitting model assumes a short (τ_1) and a long (τ_2) decay component with the corresponding amplitudes α_1 and α_2, whereas the three exponential model additionally includes a third intermediate lifetime component. A commonly used outcome parameter representing all fluorescence lifetime components is the mean fluorescence lifetime Tau mean (τ_m). τ_m can be derived from the individual lifetime components according to either of the following formula:

$$\tau_m = \frac{\tau_1 \times \alpha_1 + \tau_2 \times \alpha_2}{(\alpha_1 + \alpha_2)} \quad (2\,exponential\,fitting)$$

$$\tau_m = \frac{\tau_1 \times \alpha_1 + \tau_2 \times \alpha_2 + \tau_3 \times \alpha_3}{(\alpha_1 + \alpha_2 + \alpha_3)} \quad (3\,exponential\,fitting)$$

It represents the amplitude weighted mean fluorescence decay time per pixel and for each wavelength channel.

In either fitting approach, the first component (τ_1) usually has the largest amplitude (between 80% and 95%). It therefore also has the largest impact on τ_m. This is reasonable, as τ_1 describes the initial decline of the fluorescence, which is usually the steepest and most distinctive feature of the fluorescence decay curve. In the three-exponential approach, the third component usually has a very small amplitude (<10%), however, as it is multiplied by a relatively long τ_3 it still may have a significant influence on τ_m.

3. The decay model can be specified choosing either an incomplete or a complete decay model. For FLIO data, the incomplete model is preferred, as contribution of fluorophores with very long decay times are expected in the retina. These long-decaying molecules are still fluorescent after 12.5 ns and are therefore counted into the next decay trace.
4. The starting point of fluorescence calculation can be set to the beginning of the trace, or at the maximum of the fluorescence. Each approach features some advantages and disadvantages. The use of a tail-fit, which implies the calculation to start at the fluorescence maximum, may give an approach to eliminate the lens influence to a certain extent, as this often alters the shape of the incline within the fluorescence curve. However, this approach may also cut off some information. Therefore, it is still under discussion and may vary between individual studies.
5. Finally, there are a number of other minor adjustments that can be made, such as the number of iterations, the shift, the scatter, and the offset. These can be adjusted according to the dataset used to analyze.

In the software overview, following parameters are displayed simultaneously for the SSC and the LSC (Fig. 6.1): the fluorescence intensity window, the color coded fluorescence lifetime image of the desired parameter (mainly τ_m) with the corresponding distribution histogram with adjustable color range, the fluorescence decay curve for each pixel with corresponding decay parameters, and the chi^2 value.

FLIO measurement provides qualitative as well as quantitative data. Short fluorescence decays are depicted in red color, whereas long decays are depicted in blue. Similarly to the adjustment in brightness in FAF intensity images, color ranges in FLIO can be adjusted to highlight differences in certain areas. These adaptations are purely for visualization and do not alter the raw or fitted data. Therefore, a scale bar is typically provided beside each FLIO image to indicate the visualized range of fluorescence lifetimes.

Analysis of different decay components such as the individual components τ_1 and τ_2 with their corresponding amplitudes (α_1 and α_2) in the bi-exponential fitting approach, and τ_1, τ_2, τ_3 and α_1, α_2, α_3 in the tri-exponential fitting approach may provide further information about the distribution and contribution of individual fluorophores.

In addition, fluorescence data can be displayed and analyzed in multiple other ways, including but not limited to two dimensional graphs or mathematical approaches such as the use of a phasor transformation. Using a bi-exponential approach, the short and the long decay components can be displayed in a two dimensional histogram. Two dimensional graphs display the decay parameters of interest (e.g. τ_1 versus τ_2) and generate clouds of similar decays. Characteristic distribution clusters can be identified and assigned to specific location within the lifetime map (see also Fig. 8.1). The phasor plot works similarly, taking all lifetime decays to an imaginary space by applying a Fourier transformation and also creating clouds with similar decays. Clouds of pure fluorophores can be found along a half circle according to their decay time. Mixtures of fluorophores present as comet-like smears that are believed to connect between individual fluorophores. These clouds

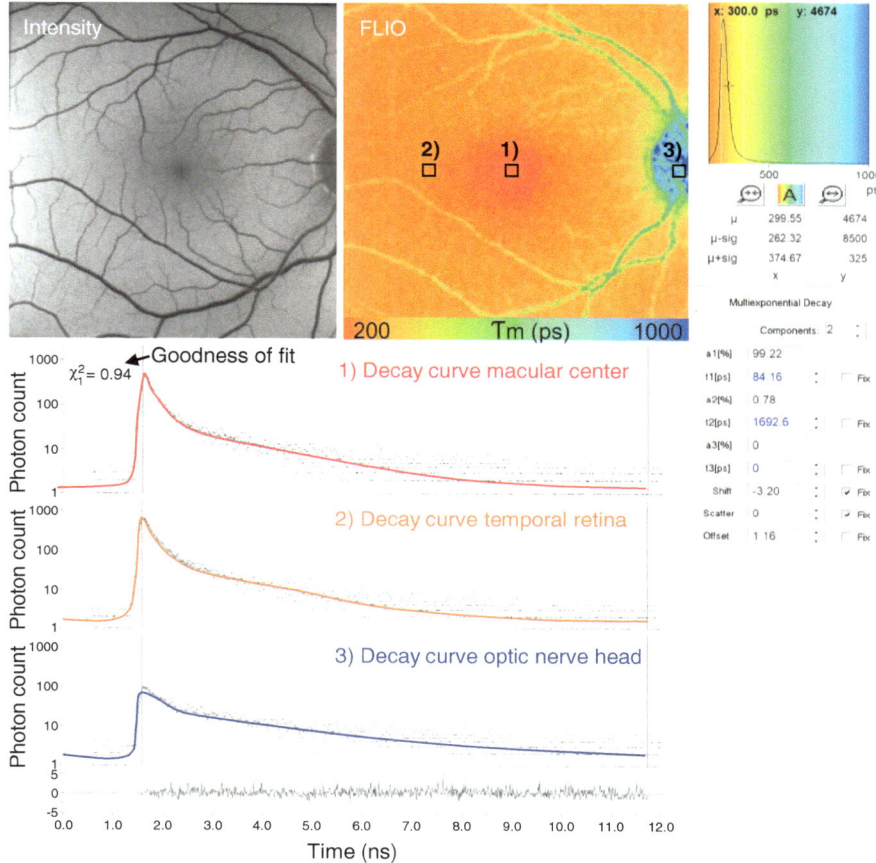

Fig. 6.1 Display and analysis of acquired FLIO data using SPCImage software. The autofluorescence intensity image and lifetime map are shown in parallel. Additionally, the distribution histogram of fluorescence lifetimes over the whole image is displayed. For each individual selected pixel, the decay parameters are provided in detail, and the goodness of the exponential fit is indicated by the x^2 value. Below, exemplary decay histograms for a pixel within the macular center (1, red), the temporal retina (2, orange), and the optic nerve head (3, blue) are displayed as the photon count over a time span of 12 ns. Right at the bottom, the residual plot is shown, which would show systematic deviation in case of inappropriate curve fitting. The shortest mean fluorescence lifetimes within the macular center derive from short decay parameters with high amplitudes. On the other hand, in the area of the optic nerve head, low amplitudes of short fluorescence lifetime parameters in combination with higher amplitudes of long decay parameters contribute to longer mean fluorescence lifetime values

may further indicate the presence of specific and potentially pathognomonic fluorophores [4].

Data as well as images of interest can then be exported for further quantification and display. For quantitative analysis of fluorescence lifetime data within individual images as well as for comparison of FLIO data between eyes, different options of software are available. Currently, the custom made FLIO reader software as well as FLIMX software can be used.

FLIO Reader

The FLIO reader (ARTORG Center for Biomedical Engineering Research, University of Bern, Bern, Switzerland) enables the display of individual lifetime- and fitting parameters including the mean fluorescence lifetime, the lifetime components τ_{1-3}, depending on a two- or three component fitting approach, the corresponding amplitudes, the photon count, and the chi^2 value (Fig. 6.2). Thereby, the selected parameter is displayed as a color coded image. A slider allows for continuous overlay with the corresponding fluorescence intensity image. An integrated standardized early treatment diabetic retinopathy study (ETDRS) grid with rings with diameters of 1 mm (center area), 3 mm (inner ring), and 6 mm (outer ring) facilitates quantification of above mentioned fluorescence lifetime parameters in the ophthalmological setting. All nine areas of the ETDRS grid can be analyzed separately. These include the center subfield, and the inner/outer ETDRS-ring, with nasal, inferior, temporal, and superior sector each. Thereby, mean fluorescence parameters and standard deviations within each grid area are calculated and displayed.

FLIMX

Another software package to analyze time resolved fluorescence data is a software package called fluorescence lifetime imaging explorer (FLIMX) [5]. Similar to SPCImage, FLIMX can apply binning to FLIO raw data and perform an exponential fitting (either an incomplete or a complete decay model) to calculate the fluorescence lifetimes from the binned data. In addition, FLIMX introduced an adaptive binning approach to minimize the loss of spatial information while ensuring a sufficient signal-to-noise ratio of the FLIO data. Furthermore, FLIMX supports advanced decay models, such as a stretched exponentials model, a layer-based approach for layered fluorescent structures, and an approach to reduce the influence of fluorescence from the crystalline lens on retinal fluorescence lifetimes. Fluorescence lifetime data from SPCImage as well as other fitting software can be imported into FLIMX for quantitative analysis and visualization. FLIO data imported into FLIMX is stored internally for documentation of the performed analysis and future evaluation. Fluorescence lifetime and fitting parameters are visualized in 1D (cross-sections) and in 2D or 3D as color-coded images with optional fluorescence intensity overlay. Fluorescence lifetimes of a single eye can be analyzed in detail and compared to FLIO data from other subjects. In order to investigate localized changes or to ensure comparability between measurements, regions

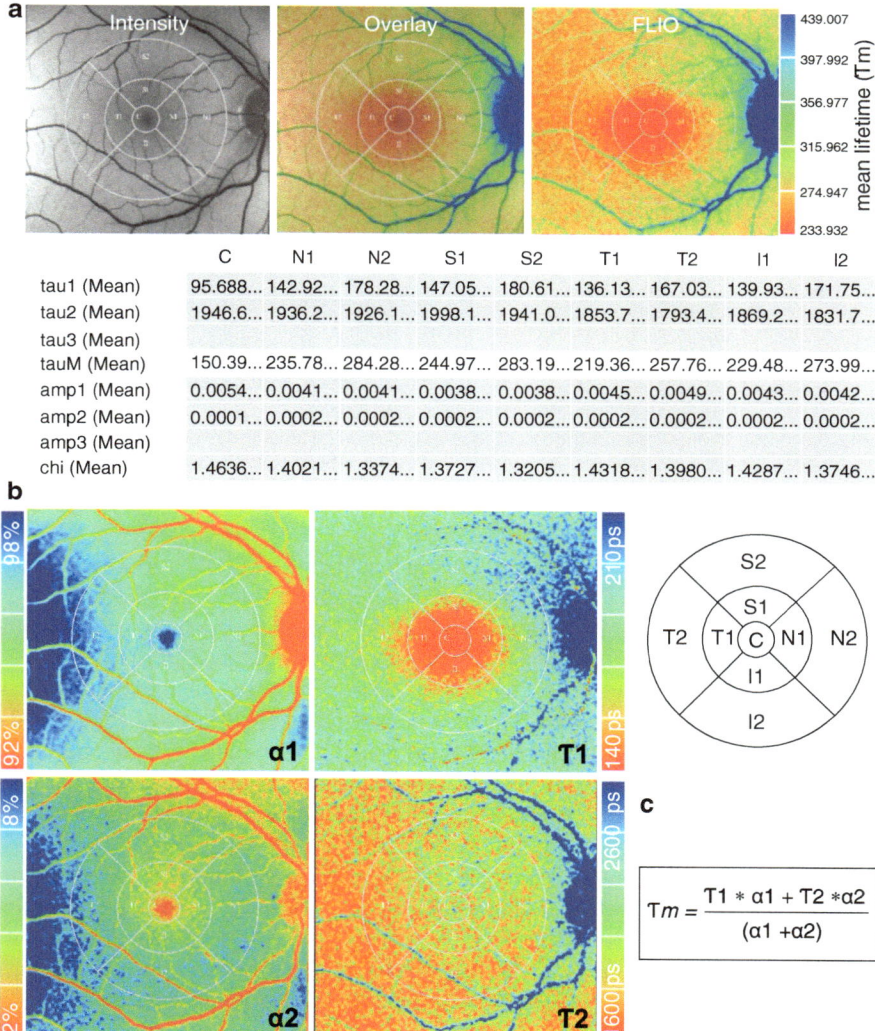

	C	N1	N2	S1	S2	T1	T2	I1	I2
tau1 (Mean)	95.688...	142.92...	178.28...	147.05...	180.61...	136.13...	167.03...	139.93...	171.75...
tau2 (Mean)	1946.6...	1936.2...	1926.1...	1998.1...	1941.0...	1853.7...	1793.4...	1869.2...	1831.7...
tau3 (Mean)									
tauM (Mean)	150.39...	235.78...	284.28...	244.97...	283.19...	219.36...	257.76...	229.48...	273.99...
amp1 (Mean)	0.0054...	0.0041...	0.0041...	0.0038...	0.0038...	0.0045...	0.0049...	0.0043...	0.0042...
amp2 (Mean)	0.0001...	0.0002...	0.0002...	0.0002...	0.0002...	0.0002...	0.0002...	0.0002...	0.0002...
amp3 (Mean)									
chi (Mean)	1.4636...	1.4021...	1.3374...	1.3727...	1.3205...	1.4318...	1.3980...	1.4287...	1.3746...

Fig. 6.2 FLIO reader software. (**a**) Calculated mean fluorescence lifetime data (τ_m) can be displayed as a continuous overlay of the fluorescence intensity image and the color coded lifetime map. A standard ETDRS grid as shown below is directly implemented and can be adjusted in location and size according to the underlying image. For each ETDRS grid area, mean values and standard deviations of the decay parameters are displayed in a table. (**b**) Color coded maps of individual decay parameters such as the amplitude α_1/α_2 and the short and long decay component τ_1/τ_2 according to the equation in (**c**) are shown as separate color coded maps

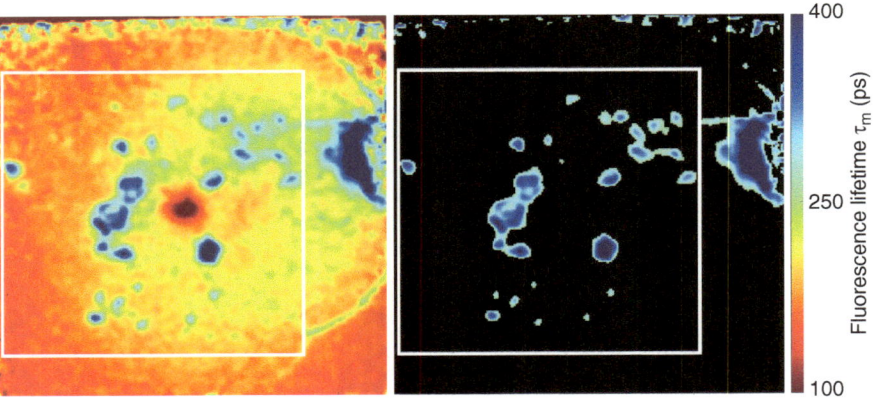

Fig. 6.3 Mean fluorescence lifetime τ_m of the short spectral channel in a patient with AMD. On the right side, fluorescence lifetimes shorter than 265 ps have been removed to segment only certain parts of the changes in an AMD patient. The color scaling is identical for both sides

of interest (ROI) can be segmented. For each subject, an ETDRS grid, two rectangles, two circles, and two free shapes can be applied. Additionally, ROIs can be combined with a threshold approach in FLIMX, as fundus structures with irregular shape or very small size can be difficult to segment manually. In Fig. 6.3, fluorescence lifetimes τ_m shorter than 265 ps have been removed to segment only certain parts of the fluorescence lifetime changes in a patient with age-related macular degeneration. A rectangular shaped ROI is applied to extract the remaining fluorescence lifetime data only from interesting parts of the fundus for further analysis, removing e.g. the optic disc from the segmented data. This method has been utilized e.g. to segment geographic atrophy regions in FLIO data [6].

For some applications, even finer grained and more sophisticated segmentation is required. Therefore, additional image data, e.g. from other imaging modalities, such as infrared images, or from advanced image processing software can be imported in FLIMX, in addition to the fluorescence lifetime data. These can be used to segment, for example, drusen in FLIO data by means of spectral unmixing [7], which would be very difficult or even impossible solely with user-defined ROIs. This is implemented by FLIMX' arithmetic images tool, which is used to calculate new image data from fluorescence lifetimes and *a priori* information and/or additionally imported image data. In addition to arithmetic operations, logical operations are also possible. Arithmetic images can process and combine data from different spectral channels and, if necessary, can be calculated on the basis of other arithmetic images, so that endless chaining is possible, which gives this tool a very large scope of application. FLIMX displays a histogram and descriptive statistics for the selected ROI and stores the

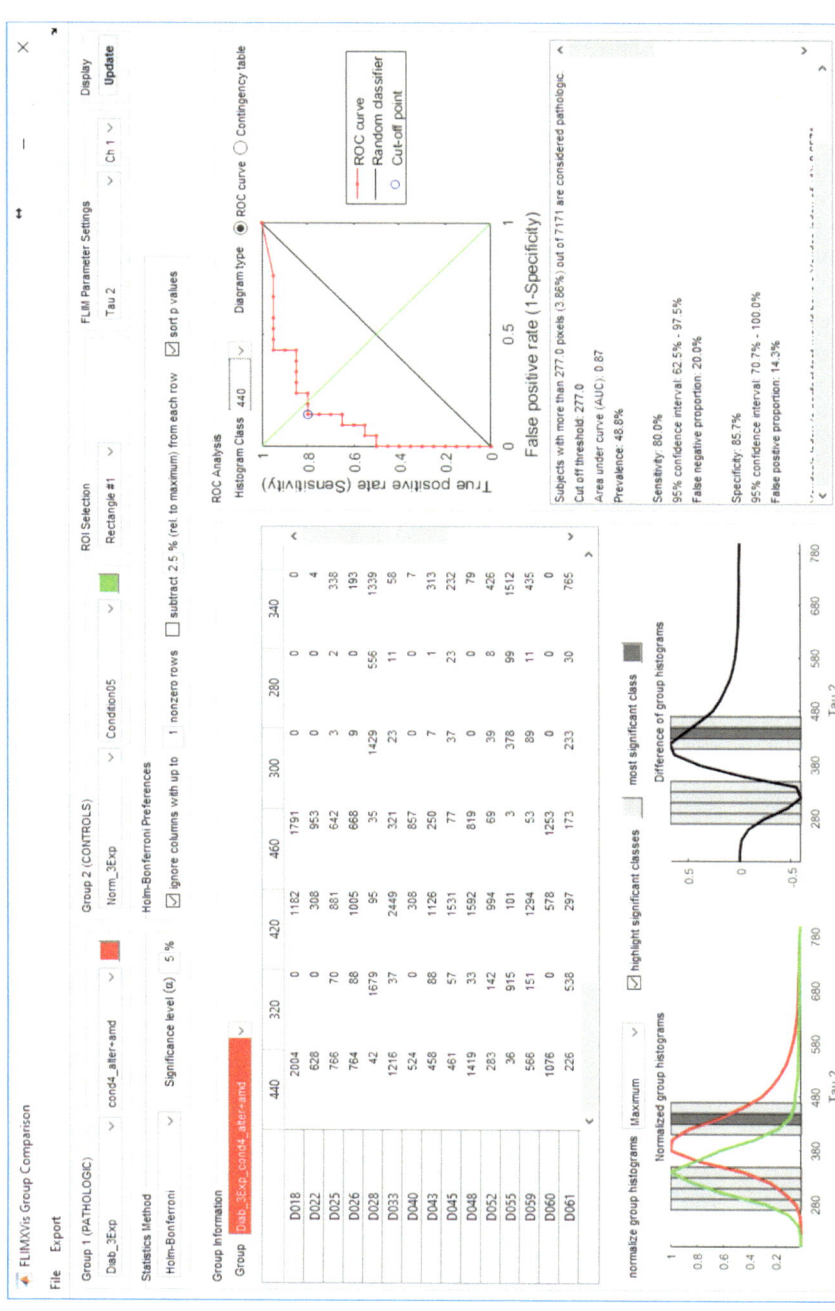

Fig. 6.4 Example of a group comparison based on FLIO data from diabetes patients and controls using the fluorescence lifetime component τ_2. The group histograms are shown in the lower left of the figure. The smallest p-value is achieved for 440 ps histogram class. The corresponding receiver operating characteristic (ROC) curve is shown on the right side

ROI coordinates as well as the thresholds and arithmetic image definitions for future use.

FLIMX allows for the analysis of subject groups using descriptive statistics and distribution tests (Lilliefors-, Shapiro-Wilk- and Kolmogorov-Smirnov-test). Subgroups can be formed according to e.g. age, disease state, etc. if that information was fed into FLIMX. A subject (sub-) group can be compared to another subject (sub-) group using a paired t-test, a two-sided t-test, a Wilcoxon signed-rang test, a Wilcoxon rank sum test or a Holm-Bonferroni method (described in [5]) as illustrated in Fig. 6.4.

The graphics generated by FLIMX can be exported in high-resolution in various file formats. Likewise, the data of the descriptive statistics as well as of the group comparison can be exported in tabular form for use in extended statistical analysis. FLIMX is documented and freely available for download online under the open source BSD–license (http://www.flimx.de).

Supplementary Material

FLIO Image Acquisition Protocol

Subjects

- Maximal dilated pupil (at least 6 mm diameter)
- Measure both eyes (internal comparison)

Measurement

Preparation

- Complete darkening of the room (switch off light, close doors, lower blinds, reduce light on computer screen)
- Move computer screen away from subjects face
- No angiography prior to FLIO (at least 2 days, better 1 week)
- No use of topical fluorescein before FLIO measurement

Before Image Acquisition

- Place subjects head correctly
- Select Tau mode on touch panel under "more"
- Use internal fixation light if possible, otherwise use external fixation light
- Standard imaging field: centered on the fovea, 30° field of view
- Optimize illumination of the fundus
- Adjust focus on small/medium sized vessels on infrared (IR) image
- After switching on the laser: acquire a single IR image, then start the measurement

During Measurement

- Check head position (forehead and chin attached to the device)
- Motivate subject to open the eyes widely, to look at the fixation target, and to keep the position of the head
- Ensure maximal and uniform illumination of the fundus
- Slight adjustment of position during measurement is possible by moving the camera head in the opposite direction of shadows in the live acquisition image: left/right, up/down; NO movement towards or away from subject
- Leave filter as deep inside as possible (deep = lower filtering); on red warning (too many photons) move filter out stepwise

End of Acquisition

- Both channels: minimum of 1000 photons in macula or at the darkest/slowest pixel (use "at cursor" function/select with mouse pointer) corresponds to approximately 1200–1500 photons in average
- Measurement time: about 1.5 to 2 min/eye in dilated pupils

References

1. Dysli C, et al. Quantitative analysis of fluorescence lifetime measurements of the macula using the fluorescence lifetime imaging ophthalmoscope in healthy subjects. Invest Ophthalmol Vis Sci. 2014;55(4):2106–13.
2. Dysli C, et al. Fluorescence lifetime imaging ophthalmoscopy. Prog Retin Eye Res. 2017;60:120–43.
3. Matthias Klemm LS, Klee S, Link D, Peters S, Hammer M, Schweitzer D, Haueisen J. Bleaching effects and fluorescence lifetime imaging ophthalmoscopy. Biomed Opt Express. 2019;10(3):1446–61.
4. Digman MA, et al. The phasor approach to fluorescence lifetime imaging analysis. Biophys J. 2008;94(2):L14–6.
5. Klemm M, et al. FLIMX: a software package to determine and analyze the fluorescence lifetime in time-resolved fluorescence data from the human eye. PLoS One. 2015;10(7):e0131640.
6. Sauer L, et al. Monitoring foveal sparing in geographic atrophy with fluorescence lifetime imaging ophthalmoscopy – a novel approach. Acta Ophthalmol. 2018;96(3):257–66.
7. Peters S, Griebsch M, Klemm M, Haueisen J, Hammer M. Hydrogen peroxide modulates energy metabolism and oxidative stress in cultures of permanent human Muller cells MIO-M1. J Biophotonics. 2017;10(9):1180–8.

Chapter 7
Fluorophores in the Eye

**Yoko Miura, Paul S. Bernstein, Chantal Dysli, Lydia Sauer,
and Martin Zinkernagel**

Introduction

There are different kinds of fluorescence molecules in the eye [1, 2]. In the intensity images of fundus autofluorescence (FAF), fluorophores with low fluorescence intensity and low tissue concentration may not be detected or outshined by fluorescence of highly concentrated strong fluorophores. In addition to characteristic excitation- and emission wavelengths, each fluorophore features an individual fluorescence lifetime. The fluorescence lifetime is specific for each molecule and independent of its concentration. Therefore, fluorescence lifetime imaging ophthalmoscopy (FLIO) may be less susceptible to the difference in concentration and fluorescent quantum yield of existing fluorophores, and has a potential to reveal weak fluorophores in FAF examinations.

For interpretation of FLIO results, fluorescence lifetime and spectral characteristics of different fluorophores within the living eye are eminently important.

Y. Miura (✉)
Institute of Biomedical Optics and Department of Ophthalmology, University of Lübeck, Lübeck, Germany
e-mail: miura@bmo.uni-luebeck.de

P. S. Bernstein
University of Utah, John A. Moran Eye Center, Salt Lake City, UT, USA

L. Sauer
Department of Ophthalmology, Moran Eye Center, University of Utah, Salt Lake City, UT, USA

C. Dysli · M. Zinkernagel
Department of Ophthalmology and Department of Clinical Research, Inselspital, Bern University Hospital, University of Bern, Bern, Switzerland

© Springer Nature Switzerland AG 2019
M. Zinkernagel, C. Dysli (eds.), *Fluorescence Lifetime Imaging Ophthalmoscopy*, https://doi.org/10.1007/978-3-030-22878-1_7

Lipofuscin (Bisretinoids)

In the visual cycle (vitamin A cycle), all-trans-retinal (atRAL) is generated in response to photoisomerization of visual pigment (e.g. rhodopsin) in photoreceptor outer segments, and its reduced form all-trans-retinol (atROL) is transported into the retinal pigment epithelial (RPE) cells. After being converted to 11-cis-retinal, it returns to the inside of the photoreceptor outer segment and binds with opsin to newly produced visual pigment (Fig. 7.1 red part) [3].

Lipofuscin is considered as a composite of at least 25 bisretinoid fluorescent substances [2], which may be formed both in photoreceptor outer segments and in RPE cells which phagocytize discarded photoreceptor outer segments [4]. Through random reactions of atRAL with the phosphatidylethanolamine (PE), a phospholipid present on the inner plate of the photoreceptor outer segment, N-retinylidene-phosphatidylethanolamine (NRPE) is formed. A second molecule of atRAL will condense with NRPE, and it may lead to the formation of dihydropyridinium-A2PE. Within the acidic and oxidizing environment, dihydropyridinium-A2PE is oxidized either to A2-dihydropyridine-PE (A2-DHP-PE) by eliminating one hydrogen or to phosphatidyl-pyridinium bisretinoid (A2PE) by eliminating two hydrogens. A2PE is either hydrolytically cleaved by enzymatic action of phosphodiesterase, which can occur in the photoreceptor outer segment before internalization by the RPE, or undergoes nonenzymatic acid-catalyzed hydrolysis inside RPE phagosomes, and N-retinylidene-N-retinylethanol-amine (A2E) is synthesized. Furthermore, various bisretinoids such as iso-A2E, iso-A2PE, the cis-trans isomers of A2E and A2PE, are also generated in this reaction process. An alternative way from NRPE goes to the formation of atRAL-dimer (atRALdi), followed by the path forming atRAL-dimer-ethanolamine (atRALdi-E) and atRAL-dimer-phospatidylethanoamine (atRALdi-PE) (Fig. 7.1 blue part) [2].

A2E was considered to be the main fluorophore of lipofuscin AF, and to undergo photooxidation that generates toxic aldehydes and ketones, impairing RPE cell function [5]. In recent studies, however, the lack of spatial correlation was proved between the distribution of A2E and lipofuscin in the human and macaque's RPE [6, 7], and thus the sources of the strong macular AF signal is still uncertain. Not only the amount, but also the photooxidation of lipofuscin bisretinoids may influence FAF, reportedly enhancing fluorescence intensity [8]. It is suggested that age-dependent RPE degeneration might be related more to the accumulation of atRALdi [9], and another report also suggest supportably that photodegradation of atRALdi might be more robust than photodegradation of A2E with respect to the contribution in fluorescence intensity [8].

Regarding the fluorescence spectrum of lipofuscin, excitation spectra of bisretinoids are slightly different from each other; A2-DHP-PE with $\lambda_{max} \sim 333$ and 490 nm, A2E with $\lambda_{max} \sim 340$ and 440 nm, atRALdi with $\lambda_{max} \sim 290$ and 432 nm, and atRALdi-PE with $\lambda_{max} \sim 333$ and 490 nm [4, 8]. While most of them show the emission spectrum around 610 nm [2, 4], atRALdi shows the emission peak around 510 nm, when excited at 430 nm [3]. It is assumed that the age-associated blue-shift

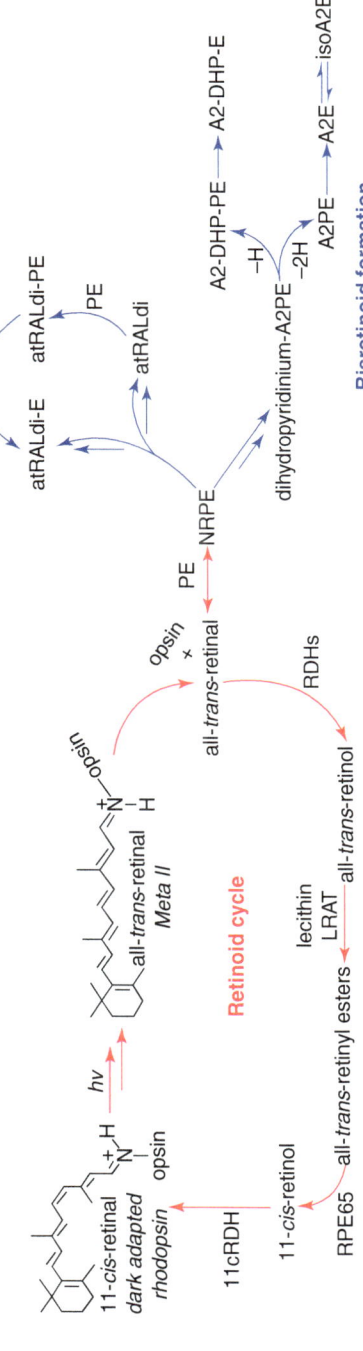

Fig. 7.1 Retinoid cycle (red) and bisretinoid formation (blue) in the eye, suggested by Sparrow et al. RDHs: retinol dehydrogenases, PE: phosphatidyletha-nolamine, RPE: retinal pigment epithelium, LRAT: lecithin retinol acyl transferase, 11cRDH: 11-cis-retinol dehydrogenase, A2-DHP-PE: A2-dihydropyridine-phosphatidylethanolamine, atRALdi: atRAL-dimer-PE, atRALdi-E: atRAL-dimer-ethanolamine. (Image Source: Sparrow et al. [3])

in the emission spectra along with the increase in AF intensity could be explained with the increase of oxidized atRALdi [8].

Detailed analysis of the time-resolved fluorescence for each lipofuscin bisretinoid is very limited and inconstant to date. For A2E the mean fluorescence lifetime is reported around 0.19 ns ($t_1 = 0.17$ ns, $\alpha_1 = 98\%$, $t_2 = 1.12$ ns, $\alpha_2 = 2\%$), whereas lipofuscin has a longer mean fluorescence lifetime of around 1.4 ns ($t_1 = 0.39$ ns, $\alpha_1 = 48\%$, $t_2 = 2.24$ ns, $\alpha_2 = 52\%$) [10]. According to the results by Cubeddu et al. the mean fluorescence lifetime of A2E solution in dimethyl sulfoxide (DMSO) and the RPE treated with A2E were 2.6 ns and 1.0 ns, respectively [11]. In another study, lipofuscin granules extracted in the chloroform from RPE cells from donor eyes showed mean fluorescence lifetime around 3 ns [12]. In clinical application of FLIO, various fluorescence lifetimes of subretinal deposits as well as their alteration over time have been presented for example in Stargardt disease, indicating a change of components of subretinal deposits during disease progression [13]. In order to make the most use of FLIO for monitoring of disease progression, further basic investigation is required to enrich the molecular background of fluorescence lifetimes in FLIO measurements.

Xantophyll (Yellow Carotinoids): Macular Pigment

Three dietary yellow carotenoids, lutein, zeaxanthin, and mesozeaxanthin (converted from lutein in the retina) are referred to as macular pigment (MP). They are concentrated in the macula lutea, accumulate primarily in the long cone receptor axons (Henle's fibers) in the fovea. MP can also be found in the inner plexiform layer in the parafoveal area [14, 15]. Lutein and zeaxanthin in plasma are taken up from the choriocapillary layer via RPE and probably also via Müller cells into the photoreceptor outer segments [16, 17], and accumulate in the axons of photoreceptors as well as in the inner plexiform layer. For intraretinal uptake, carotenoid binding proteins are suggested as the transporter of these carotenoids [18]. MP plays important roles in protecting the retina from photochemical damage through absorbing high energy short-wavelength visible light (blue light). Furthermore, MP has been shown to have an antioxidant function [19].

Positive role of MP in retinal health has been further proven in the last years, particularly their potential in protection from age-related macular degeneration [16, 20, 21]. Dietary uptake of lutein or zeaxanthin has been shown to increase their density in the human retina [22, 23], and thus the efficacy of supplementation is actively discussed and studied today.

The absorption maximum of MP is around 450 to 480 nm, the spectrum covers the blue light wavelength range [24]. The emission maximum is in the range of orange-red around 600–700 nm. Due to its low fluorescence efficiency, MP shows hypofluorescence in FAF examination using short/blue wavelength excitation compared to other retinal fluorophores. However, MP was shown to feature very short mean fluorescence lifetime (<0.1 ns) [25]. In FLIO, probably due to the fact that the fluorescence lifetime is independent of the fluorescence intensity, the contribution of MP in the fluorescence lifetime at the fovea is shown to be significant [25, 26].

Collagen/Elastin

Collagen is the main structural protein in the extracellular space in various connective tissues. In the human fundus, collagen can be found as a component of vessels in the retina and choroid, Bruch's membrane, and the vitreous. In human retinal vessels, collagen types I, III, and IV–VI were found in large vessels, types I, IV, and V plus small amounts of III and VI in small vessels. The RPE basal lamina contains collagen types IV and V. The inner and outer collagenous layer of the Bruch's membrane contains fibers of collagens types I, III, and V [27]. The choriocapillaris basal lamina is similar to the RPE basal lamina, but contains intense amount of types VI collagen additionally. In the choroidal stroma, type IV collagen is also localized at all basement membranes [28]. An increase of type IV collagen is suggested to contribute to the age-related thickening of Bruch's membrane [27].

The vitreous body is dominated by collagen type II [29], but also contains other collagen types (II, V, IX, and XI) [30]. At the vitreoretinal interface, collagen types II, IV, VI, and XVIII are positive [30], whereby type VI collagen is considered to be most essential in the organization of the vitreous fibers into lamellae penetrating the internal limiting membrane [31]. The attachment of the vitreous collagen to optic disc is generally stronger than the other retinal surfaces.

Elastin is also one of the components of connective tissue, and can be found in vessels, or basal membranes. Elastin is contained more in arteries than in veins [32]. The elastic fiber layer in the Bruch's membrane is a 0.8-μm-thick sheet, and also contains collagen VI and other extracellular matrices, and is connected to the collagen fibers in the inner and outer collagen layers [33].

Fluorescence properties of collagen and elastin have been studied mainly for ultraviolet fluorescence excited by the light with wavelength shorter than 400 μm [34, 35]. However, collagens and elastin are also fluorescent, although weak, under excitation at 450–480 nm, and the emission maximum lays around 470–520 nm [10]. With respect to the time-resolved fluorescence property of collagen, Schweitzer et al. showed that different types of collagens show different fluorescence lifetimes, where collagen type I, II, III, and IV had a mean lifetime value of 1.75 ns, 1.44 ns, 1.11 ns, and 1.62 ns, respectively [10]. The fluorescence lifetime of elastin has been reported with 1.38 ns [10].

Redox Coenzyme

Redox coenzymes in cells, such as flavin adenine dinucleotide (FAD), flavin mononucleotide (FMN), and nicotinamide adenine dinucleotide (NADH), are fluorescent. These coenzymes are involved in several important enzymatic reactions in cell energy metabolism.

FAD (the oxidized form of FADH or $FADH_2$) and FMN (the oxidized form of $FMNH_2$) derive from riboflavin, vitamin B2, and function as cofactors in many reactions of intermediary metabolism, such as carbohydrate, fat, and amino acid synthesis. They are also required for the activation of other vitamins such as vitamin B6 (Table 7.1). FMN is involved in the first enzyme in the complex I of the electron transport chain in mitochondria, where it acts as an electron carrier to oxidize NADH to NAD. FAD is used in a many cellular reactions, among which it acts as a coenzyme of succinate dehydrogenase in the TCA cycle (Krebs cycle) that is at the same time the complex II reaction of electron transport chain in mitochondrial inner membrane. The enzymes bound to FAD has an individual FAD-binding site [36].

NADH is synthesized from tryptophan. NADH is a reduced form and acts also as a cofactor. In cell metabolic reactions, it acts as an electron acceptor, and is oxidized to NAD. NADH is required for many reactions such as beta oxidation, glycolysis, TCA cycle, and electron transport chain reactions (Table 7.2). The NADH formed in the TCA cycle is utilized as the first step of electron transport chain reactions, where NADH acts as a carrier of electron to the complex I. This electron is transferred further to produce ATP.

These co-factors have interesting fluorescence properties. Regarding flavins, FAD or FMN, the oxidized form, is fluorescent, whereas their reduced forms (FADH and FADH2) show very weak or almost no fluorescence [37]. Their maxima of λ_{ex}/λ_{em} are reported to be 370 nm and 450 nm/535 nm [38]. Previous studies have shown that free FAD has a long fluorescence lifetime (2.3 ± 0.7 ns) while the protein-bound FAD has a significantly shorter fluorescence lifetime (0.13 ± 0.02 ns) for the monomeric form, and 0.04 ± 0.01 ns for the dimeric form [39]. Cellular metabolic changes associated with the shift in energy supply between the aerobic and anaerobic

Table 7.1 Enzymes bound to FAD/FMN (partial)

Enzyme	Cofactor	Function
Succinate dehydrogenase	FAD	Krebs cycle/electron transport chain complex II
Dihydrolipoyl dehydrogenase	FAD	Energy metabolism as the part of pyruvate dehydrogenase complex
Fatty acyl-CoA dehydrogenase	FAD	Fatty acid oxidation
Glutathione reductase	FAD	Reduction of GSSG (oxidized glutathione)
NADH dehydrogenase	FMN	Electron transport chain complex I
Pyridoxine phosphate oxidase	FMN	Vitamin B6 metabolism

Table 7.2 Enzymes bound to NADH (partial)

Enzyme	Function
Pyruvate dehydrogenase	Formation from pyruvate to acetyl-CoA
Malate dehydrogenase	Krebs cycle
Alpha-ketoglutarate dehydrogenase	Krebs cycle
Isocitrate dehydrogenase	Krebs cycle
Glyceraldehyde-3-phosphate dehydrogenase	Glycolysis
3-Hydroxyacyl-CoA dehydrogenase	Beta oxidation

metabolism, for example through the change in oxygen supply or mitochondrial activity, could significantly alter the protein bound-to-free form ratio of the FAD, which may consequently influence the fluorescence lifetime. There are many basic researches demonstrating the utility of the fluorescence lifetime measurement of FAD to assess energy metabolic status in cells and tissues [40, 41]. Regarding FMN, the fluorescence lifetime in its free form is reportedly 4.7 ns [42]. No literatures for the fluorescence lifetime of the bound form of FMN in the form of co-factor has been presented to date. Since the fluorescence spectrum of FAD and FMN is within the FLIO detection range, it is well assumed that FAD and FMN may also contribute to the retinal fluorescence lifetime in FLIO.

On the other hand, the reduced form NADH is fluorescent whereas the oxidized form NAD is not. NADH has also different fluorescence lifetimes according to its protein-binding states; different from FAD, the free NADH has a short fluorescence lifetime (0.30 ± 0.03 ns), and the protein-bound NADH shows a longer one (2.44 ± 0.06 ns) [40, 43]. Using UV-excitation or two-photon excitation, these properties of NADH fluorescence lifetime have been utilized widely to assess cell metabolic states [40, 44, 45]. However, as the fluorescence spectrum of NADH lays outside of the spectral range of FLIO, it is unlikely that NADH-AF and its fluorescence lifetime may be directly measured with FLIO.

Melanin/Oxidized Melanin

In recent years, increasing evidence in FAF examination has shown that near-infrared AF may indicate melanin distribution within the fundus [46, 47]. Although melanin has a high absorption coefficient over wide range of wavelengths from UV to infrared, its quantum efficiency is quite low in the range of visual light [48, 49]. It results in hypofluorescence in FAF examination with short wavelength excitation. However, the explanation for the melanin AF in the near-infrared FAF examination is still under discussion.

The emission maximum of melanin is shown around 540 nm when excited with 470 nm [50]. With respect to the fluorescence lifetime of melanin in RPE cells, laboratory works using ex-vivo RPE tissues showed short mean fluorescence lifetimes of melanosomes around 0.1–0.2 ns [51, 52]. Measurement of nevi and melanocytes also presents similar short fluorescence lifetime as measured with the melanosomes in ex-vivo RPE [53]. In a study with synthetic melanin in DMSO and phosphate-buffered saline (PBS) the melanin fluorescence lifetime was 0.84 ns and 0.9 ns, respectively [50]. In the same study, the melanin incubated with hydrogen peroxide (oxidized melanin) showed an extension of the fluorescence lifetime, almost factor 2 (1.75 ns). The interpretation of the authors was that this was caused by the enhancement of the fluorescence quantum yield of the melanin. As mechanism behind, oxidation may lead to the degradation of melanin polymerization that causes fluorescence quenching, and thus it resulted in the enhancement of its quantum yield. This result coincides with another study conducted with human donor eye and synthetic melanin, showing that oxidation increased the intensity of melanin AF in the range of visual wavelength [54].

The detection of short fluorescence lifetimes of the RPE melanin with FLIO has also been confirmed experimentally by the author. The RPE from young porcine eyes (without lipofuscin, abundant melanosomes at the apical side) placed in the artificial eye chamber filled with PBS showed mean fluorescence lifetime of 0.24 ± 40 ns (short spectral channel: SSC) and 0.19 ± 20 ns (long spectral channel: LSC) [55], which are similar to the ones obtained from microscopic studies described above. The exemplary images of ex-vivo FLIO, examining different fundus layers of a porcine eye, are shown in Fig. 7.2.

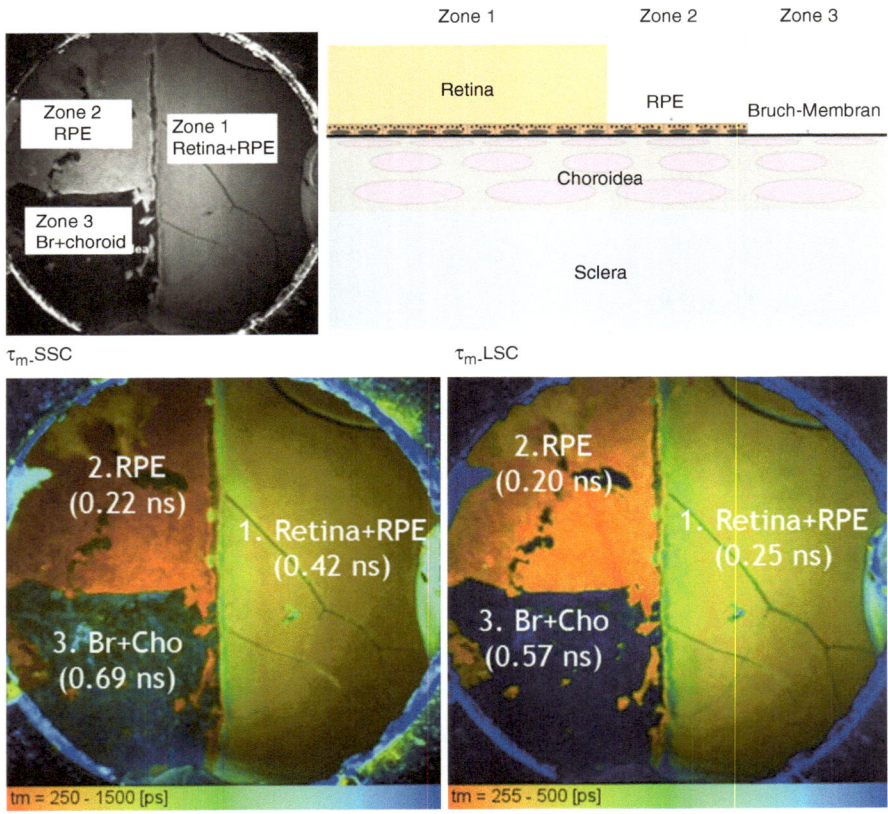

Fig. 7.2 FLIO with ex-vivo porcine fundus tissues. The posterior part of the porcine eye was carefully separated in three parts; (1) all layers (retina + RPE (+Bruch's membrane (Br) + choroid (Cho) behind)), (2) without retina (only RPE (+Br + Cho behind)), (3) only Br + Cho (the RPE was removed with a fine brush). The RPE without or very little amount of lipofuscin showed short mean fluorescence lifetime around 0.20–0.25 ns. It is assumed to be mainly originated from melanosomes and coincides to the results with previous microscopic experiments

Lipid Peroxidation Endproducts

In 1969, Chio et al. reported that the lipid peroxidation of unsaturated fatty acids produces fluorescent products similar to lipofuscin [56], and since then the theory has been supported for many years that lipofuscin is the endproduct of lipid peroxidation of unsaturated fatty acids in the phagocytized photoreceptor outer segments. However, it was refuted later, due to the fact that the fluorescence spectrum of lipid peroxidation products ($\lambda_{ex}/\lambda_{em} \approx 360$ nm/515 nm) turned out to differ from the ones of lipofuscin ($\lambda_{ex}/\lambda_{em} \approx 420$ nm/600 nm) [57], suggesting that it is most likely true that the fluorescent substances observed in the clinical FAF examination are different from lipid peroxidation endproducts, but it could be the degradative byproduct of retinol cycle, bisretinoids. The latter explanation is widely accepted today. However, it is not ignorable, that many further in vivo and ex vivo experiments reported the appearance of lipofuscin-like fluorescent substances in/around RPE cells under oxidative stress, which show a fluorescence spectrum that is different from the one of lipofuscin [51, 58, 59]. These lipofuscin-like fluorescent substances are assumed to be phagosomes and/or the storage granule of retinyl ester in/around RPE cells that underwent lipid peroxidation [51, 60]. Although their fluorescence spectrum is not exactly same as the one of lipofuscin, the linkage of these granules to the lipofuscin formation was still not elucidated yet. Fluorescence lifetime of these high fluorescent granules was significantly longer than the one of untreated RPE (Fig. 7.3) [51].

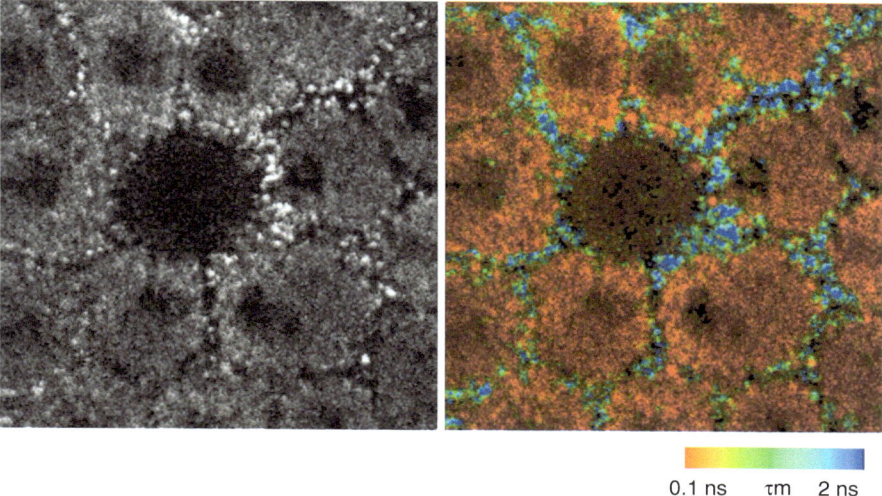

0.1 ns   τm   2 ns

Fig. 7.3 Two-photon microscopy with fluorescence lifetime imaging with ex-vivo porcine RPE incubated with ferrous sulfate (FeSO$_4$) (strong inducer of lipid peroxidation) for 1 h. Bright granular autofluorescence with long fluorescenece lifetime appeared in/around the PPE cells. They are considered the retinol cylce-related lipid that underwent peroxidation. $\lambda_{ex} = 750$ nm (corresponds to 375 nm in one photon excitation), emission range is from 380 nm to 680 mm. The emission maximum was found around 500 nm. These fluorescence spectral properties fit very well to the previously reported ones of lipid peroxidation producs

Advanced Glycation Endproducts (AGE)

Formation of AGEs results from the nonenzymatic reaction to bond sugar molecule, such as glucose or fructose, to proteins, lipids, and nucleic acids. Some AGEs are very reactive and may damage cells and tissues directly, and thus are crucial in ageing and in different diseases. In ophthalmology, AGE is known to be implicated in the pathogenesis of different retinal diseases [61, 62], especially in diabetic retinopathy (DR) [61]. AGEs accumulate in the retina in DR, for example in retinal pericytes, mediating pericyte survival and function that consequently leads to the pericyte loss [63]. The functional damage of Müller cells is also suggested [64]. Not only in the retina, AGE amount may increase also in the crystalline lens [65] and in the cornea [66] in diabetic patients.

Typical fluorescence spectra of AGE is $\lambda_{ex}/\lambda_{em} \approx 370\,nm/440\,nm$ [67]. According to the measurement by Schweitzer et al., AGE shows an emission maximum around 490 nm when excited with 468 nm [10]. Reportedly, mean fluorescence lifetime of AGE is 1.7 ns [10]. In diabetic patients, fluorescence lifetimes may be increased significantly [68], considered to be due to the accumulation of AGE in the lens and the neural retina, as well as the increase of the relative amount of free form of FAD [68].

Other Fluorophores

Additionally to the fluorophores described above, retinol, aromatic amino acids like tryptophan, tyrosine, and phenylalanine, and pyridoxine (vitamin B6) are known to be fluorescent. They all have fluorescence spectra outside the excitation and/or detection range of FLIO, and therefore it is unlikely that their fluorescence lifetimes can be detected and contribute significantly to FLIO. Briefly, the fluorescence excitation/emission maxima of retinol are $\lambda_{ex}/\lambda_{em} \approx 330\,nm/474\,nm$ [69]. Regarding aromatic amino acids, they are 280 nm/350 nm, 274 nm/303 nm, and 274 nm/303 nm, for tryptophan, tyrosine, and phenylalanine, respectively [70]. Pyridoxine has a fluorescence maximum at $\lambda_{ex}/\lambda_{em} \approx 296\,nm/396\,nm$ [71].

References

1. Delori FC, Dorey CK, Staurenghi G, Arend O, Goger DG, Weiter JJ. In vivo fluorescence of the ocular fundus exhibits retinal pigment epithelium lipofuscin characteristics. Invest Ophthalmol Vis Sci. 1995;36:718–29.
2. Sparrow JR, Gregory-Roberts E, Yamamoto K, et al. The bisretinoids of retinal pigment epithelium. Prog Retin Eye Res. 2012;31:121–35.
3. Sparrow JR, Wu YL, Kim CY, Zhou JL. Phospholipid meets all-trans-retinal: the making of RPE bisretinoids. J Lipid Res. 2010;51:247–61.

4. Sparrow JR, Wu Y, Nagasaki T, Yoon KD, Yamamoto K, Zhou J. Fundus autofluorescence and the bisretinoids of retina. Photochem Photobiol Sci. 2010;9:1480–9.
5. Wang Z, Keller LM, Dillon J, Gaillard ER. Oxidation of A2E results in the formation of highly reactive aldehydes and ketones. Photochem Photobiol. 2006;82:1251–7.
6. Ablonczy Z, Higbee D, Anderson DM, et al. Lack of correlation between the spatial distribution of A2E and lipofuscin fluorescence in the human retinal pigment epithelium. Invest Ophthalmol Vis Sci. 2013;54:5535–42.
7. Pallitto P, Ablonczy Z, Jones EE, et al. A2E and lipofuscin distributions in macaque retinal pigment epithelium are similar to human. Photochem Photobiol Sci. 2015;14:1888–95.
8. Kim SR, Jang YP, Sparrow JR. Photooxidation of RPE lipofuscin bisretinoids enhances fluorescence intensity. Vis Res. 2010;50:729–36.
9. Zhao J, Liao Y, Chen J, et al. Aberrant buildup of all-trans-retinal dimer, a nonpyridinium bisretinoid lipofuscin fluorophore, contributes to the degeneration of the retinal pigment epithelium. Invest Ophthalmol Vis Sci. 2017;58:1063–75.
10. Schweitzer D, Schenke S, Hammer M, et al. Towards metabolic mapping of the human retina. Microsc Res Tech. 2007;70:410–9.
11. Cubeddu R, Taroni P, Hu DN, Sakai N, Nakanishi K, Roberts JE. Photophysical studies of A2-E, putative precursor of lipofuscin, in human retinal pigment epithelial cells. Photochem Photobiol. 1999;70:172–5.
12. Yakovleva MA, Feldman TB, Arbukhanova PM, Borzenok SA, Kuzmin VA, Ostrovsky MA. The fluorescence lifetime of lipofuscin granule fluorophores contained in the retinal pigment epithelium cells from human cadaver eyes in normal state and in the case of visualized pathology. Dokl Biochem Biophys. 2017;474:239–43.
13. Dysli C, Wolf S, Hatz K, Zinkernagel MS. Fluorescence lifetime imaging in Stargardt disease: potential marker for disease progression. Invest Ophthalmol Vis Sci. 2016;57:832–41.
14. Bone RA, Landrum JT. Distribution of macular pigment components, zeaxanthin and lutein, in human retina. Methods Enzymol. 1992;213:360–6.
15. Snodderly DM, Auran JD, Delori FC. The macular pigment. II. Spatial distribution in primate retinas. Invest Ophthalmol Vis Sci. 1984;25:674–85.
16. During A, Doraiswamy S, Harrison EH. Xanthophylls are preferentially taken up compared with beta-carotene by retinal cells via a SRBI-dependent mechanism. J Lipid Res. 2008;49:1715–24.
17. Tserentsoodol N, Gordiyenko N, Pascual I, Lee J, Fliesler S, Rodriguez I. Intraretinal lipid transport is dependent on high density lipoprotein-like particles and class B scavenger receptors. Mol Vis. 2006;12:1319–33.
18. Arunkumar R, Calvo CM, Conrady CD, Bernstein PS. What do we know about the macular pigment in AMD: the past, the present, and the future. Eye. 2018;32:992–1004.
19. Kamoshita M, Toda E, Osada H, et al. Lutein acts via multiple antioxidant pathways in the photo-stressed retina. Sci Rep. 2016;6:30226.
20. Obana A, Hiramitsu T, Gohto Y, et al. Macular carotenoid levels of normal subjects and age-related maculopathy patients in a Japanese population. Ophthalmology. 2008;115:147–57.
21. Meyers KJ, Mares JA, Igo RP, et al. Genetic evidence for role of carotenoids in age-related macular degeneration in the carotenoids in age-related eye disease study (CAREDS). Invest Ophthalmol Vis Sci. 2014;55:587–99.
22. Corvi F, Souied EH, Falfoul Y, et al. Pilot evaluation of short-term changes in macular pigment and retinal sensitivity in different phenotypes of early age-related macular degeneration after carotenoid supplementation. Br J Ophthalmol. 2017;101:770–3.
23. Bernstein PS, Li B, Vachali PP, et al. Lutein, zeaxanthin, and meso-zeaxanthin: the basic and clinical science underlying carotenoid-based nutritional interventions against ocular disease. Prog Retin Eye Res. 2016;50:34–66.
24. Bernstein PS, Delori FC, Richer S, van Kuijk FJ, Wenzel AJ. The value of measurement of macular carotenoid pigment optical densities and distributions in age-related macular degeneration and other retinal disorders. Vis Res. 2010;50:716–28.

25. Sauer L, Andersen KM, Li B, Gensure RH, Hammer M, Bernstein PS. Fluorescence lifetime imaging ophthalmoscopy (FLIO) of macular pigment. Invest Ophthalmol Vis Sci. 2018;59:3094–103.
26. Sauer L, Schweitzer D, Ramm L, Augsten R, Hammer M, Peters S. Impact of macular pigment on fundus autofluorescence lifetimes. Invest Ophthalmol Vis Sci. 2015;56:4668–79.
27. Marshall GE, Konstas AG, Reid GG, Edwards JG, Lee WR. Collagens in the aged human macula. Graefes Arch Clin Exp Ophthalmol. 1994;232:133–40.
28. Marshall GE, Konstas AG, Reid GG, Edwards JG, Lee WR. Type IV collagen and laminin in Bruch's membrane and basal linear deposit in the human macula. Br J Ophthalmol. 1992;76:607–14.
29. van Deemter M, Kuijer R, Harm Pas H, Jacoba van der Worp R, Hooymans JM, Los LI. Trypsin-mediated enzymatic degradation of type II collagen in the human vitreous. Mol Vis. 2013;19:1591–9.
30. Ponsioen TL, van Luyn MJ, van der Worp RJ, van Meurs JC, Hooymans JM, Los LI. Collagen distribution in the human vitreoretinal interface. Invest Ophthalmol Vis Sci. 2008;49:4089–95.
31. Bu SC, Kuijer R, van der Worp RJ, Li XR, Hooymans JMM, Los LI. The ultrastructural localization of type II, IV, and VI collagens at the vitreoretinal interface. PLoS One. 2015;10:e0134325.
32. Basu P, Sen U, Tyagi N, Tyagi SC. Blood flow interplays with elastin: collagen and MMP: TIMP ratios to maintain healthy vascular structure and function. Vasc Health Risk Manag. 2010;6:215–28.
33. Korte GE, D'Aversa G. The elastic tissue of Bruch's membrane. Connections to choroidal elastic tissue and the ciliary epithelium of the rabbit and human eyes. Arch Ophthalmol. 1989;107:1654–8.
34. Schenke-Layland K. Non-invasive multiphoton imaging of extracellular matrix structures. J Biophotonics. 2008;1:451–62.
35. Blomfield J, Farrar JF. The fluorescent properties of maturing arterial elastin. Cardiovasc Res. 1969;3:161–70.
36. Cheng VW, Piragasam RS, Rothery RA, Maklashina E, Cecchini G, Weiner JH. Redox state of flavin adenine dinucleotide drives substrate binding and product release in Escherichia coli succinate dehydrogenase. Biochemistry. 2015;54:1043–52.
37. Galban J, Sanz-Vicente I, Navarro J, de Marcos S. The intrinsic fluorescence of FAD and its application in analytical chemistry: a review. Methods Appl Fluoresc. 2016;4:042005.
38. Kotaki A, Yagi K. Fluorescence properties of flavins in various solvents. J Biochem. 1970;68:509–16.
39. Nakashima N, Yoshihara K, Tanaka F, Yagi K. Picosecond fluorescence lifetime of the coenzyme of D-amino-acid oxidase. J Biol Chem. 1980;255:5261–3.
40. Skala MC, Riching KM, Gendron-Fitzpatrick A, et al. In vivo multiphoton microscopy of NADH and FAD redox states, fluorescence lifetimes, and cellular morphology in precancerous epithelia. Proc Natl Acad Sci U S A. 2007;104:19494–9.
41. Chakraborty S, Nian FS, Tsai JW, Karmenyan A, Chiou A. Quantification of the metabolic state in cell-model of Parkinson's disease by fluorescence lifetime imaging microscopy. Sci Rep. 2016;6:19145.
42. Wahl P, Auchet JC, Visser AJ, Muller F. Time resolved fluorescence of flavin adenine dinucleotide. FEBS Lett. 1974;44:67–70.
43. Lakowicz JR, Szmacinski H, Nowaczyk K, Johnson ML. Fluorescence lifetime imaging of free and protein-bound Nadh. Proc Natl Acad Sci U S A. 1992;89:1271–5.
44. Szaszak M, Chang JC, Leng W, Rupp J, Ojcius DM, Kelley AM. Characterizing the intracellular distribution of metabolites in intact chlamydia-infected cells by Raman and two-photon microscopy. Microbes Infect. 2013;15:461–9.
45. Cong A, Pimenta RML, Lee HB, Mereddy V, Holy J, Heikal AA. Two-photon fluorescence lifetime imaging of intrinsic NADH in three-dimensional tumor models. Cytometry A. 2019;95(1):80–92.

46. Oguchi Y, Sekiryu T, Takasumi M, Hashimoto Y, Furuta M. Near-infrared and short-wave autofluorescence in ocular specimens. Jpn J Ophthalmol. 2018;62:605–13.
47. Paavo M, Zhao J, Kim HJ, et al. Mutations in GPR143/OA1 and ABCA4 inform interpretations of short-wavelength and near-infrared fundus autofluorescence. Invest Ophthalmol Vis Sci. 2018;59:2459–69.
48. Meredith P, Riesz J. Radiative relaxation quantum yields for synthetic eumelanin. Photochem Photobiol. 2004;79:211–6.
49. Nighswander-Rempel SP, Riesz J, Gilmore J, Meredith P. A quantum yield map for synthetic eumelanin. J Chem Phys. 2005;123:194901.
50. Colbert A, Scholar M, Heikal AA. Towards probing skin Cancer using endogenous melanin fluorescence. Penn State McNair J. 2004;11:8–15.
51. Miura Y, Huettmann G, Orzekowsky-Schroeder R, et al. Two-photon microscopy and fluorescence lifetime imaging of retinal pigment epithelial cells under oxidative stress. Invest Ophthalmol Vis Sci. 2013;54:3366–77.
52. Hammer M, Sauer L, Klemm M, Peters S, Schultz R, Haueisen J. Fundus autofluorescence beyond lipofuscin: lesson learned from ex vivo fluorescence lifetime imaging in porcine eyes. Biomed Opt Express. 2018;9:3078–91.
53. Dimitrow E, Riemann I, Ehlers A, et al. Spectral fluorescence lifetime detection and selective melanin imaging by multiphoton laser tomography for melanoma diagnosis. Exp Dermatol. 2009;18:509–15.
54. Kayatz P, Thumann G, Luther TT, et al. Oxidation causes melanin fluorescence. Invest Ophthalmol Vis Sci. 2001;42:241–6.
55. Miura Y, Lewke B, Hutfilz A, Brinkmann R. Change in fluorescence lifetime of retinal pigment epithelium under oxidative stress. Nippon Ganka Gakkai Zasshi. 2019;123:105–14.
56. Chio KS, Reiss U, Fletcher B, Tappel AL. Peroxidation of subcellular organelles: formation of lipofuscinlike fluorescent pigments. Science. 1969;166:1535–6.
57. Eldred GE, Katz ML. The autofluorescent products of lipid peroxidation may not be lipofuscin-like. Free Radic Biol Med. 1989;7:157–63.
58. Katz ML, Christianson JS, Gao CL, Handelman GJ. Iron-induced fluorescence in the retina: dependence on vitamin a. Invest Ophthalmol Vis Sci. 1994;35:3613–24.
59. Krohne TU, Stratmann NK, Kopitz J, Holz FG. Effects of lipid peroxidation products on lipofuscinogenesis and autophagy in human retinal pigment epithelial cells. Exp Eye Res. 2010;90:465–71.
60. Imanishi Y, Batten ML, Piston DW, Baehr W, Palczewski K. Noninvasive two-photon imaging reveals retinyl ester storage structures in the eye. J Cell Biol. 2004;164:373–83.
61. Xu J, Chen LJ, Yu J, et al. Involvement of advanced Glycation end products in the pathogenesis of diabetic retinopathy. Cell Physiol Biochem. 2018;48:705–17.
62. Banevicius M, Vilkeviciute A, Kriauciuniene L, Liutkeviciene R, Deltuva VP. The association between variants of receptor for advanced Glycation end products (RAGE) gene polymorphisms and age-related macular degeneration. Med Sci Monit. 2018;24:190–9.
63. Hammes HP, Lin J, Renner O, et al. Pericytes and the pathogenesis of diabetic retinopathy. Diabetes. 2002;51:3107–12.
64. Thompson K, Chen J, Luo QY, Xiao YC, Cummins TR, Bhatwadekar AD. Advanced glycation end (AGE) product modification of laminin downregulates Kir4.1 in retinal Muller cells. PLoS One. 2018;13:e0193280.
65. Abiko T, Abiko A, Ishiko S, Takeda M, Horiuchi S, Yoshida A. Relationship between autofluorescence and advanced glycation end products in diabetic lenses. Exp Eye Res. 1999;68:361–6.
66. Sato E, Mori F, Igarashi S, et al. Corneal advanced glycation end products increase in patients with proliferative diabetic retinopathy. Diabetes Care. 2001;24:479–82.
67. Yanagisawa K, Makita Z, Shiroshita K, et al. Specific fluorescence assay for advanced glycation end products in blood and urine of diabetic patients. Metabolism. 1998;47:1348–53.
68. Schweitzer D, Deutsch L, Klemm M, et al. Fluorescence lifetime imaging ophthalmoscopy in type 2 diabetic patients who have no signs of diabetic retinopathy. J Biomed Opt. 2015;20:061106.

69. Torre M, San Andres MP, Vera S, Montalvo G, Valiente M. Retinol fluorescence: a simple method to differentiate different bilayer morphologies. Colloid Polym Sci. 2009;287:951–9.
70. Ghisaidoobe ABT, Chung SJ. Intrinsic tryptophan fluorescence in the detection and analysis of proteins: a focus on Forster resonance energy transfer techniques. Int J Mol Sci. 2014;15:22518–38.
71. Algar SO, Martos NR, Diaz AM. Native fluorescence determination of pyridoxine hydrochloride (vitamin B-6) in pharmaceutical preparations after sorption on sephadex SP C-25. Spectrosc Lett. 2003;36:133–49.

Chapter 8
FLIO in the Healthy Eye

Chantal Dysli and Muriel Dysli

In order to understand FLIO, and to compare and quantify fluorescence lifetimes in diseased retinas, FLIO data of healthy eyes is required. Conventional FAF intensity imaging reveals characteristic distribution patterns of high and low autofluorescence in healthy eyes [1]. FAF intensity primarily reflects the fluorescence signal of a dominant retinal fluorophore, likely lipofuscin, and is mainly originating from the RPE [1]. Further contributions of minor fluorophores from the outer retinal layers and the photoreceptors are possible as well [1]. Dominant fluorophores lead to bright areas in FAF intensity images. On the other hand, retinal vessels and the optic nerve head naturally feature very low autofluorescence intensity, and thus appear dark in FAF intensity images. The foveal center also presents with low autofluorescence intensities due to the absorption of both, excitation light as well as RPE fluorescence trough the macular pigment (MP) [2]. MP will be further described in Chap. 16.

Analogous to FAF intensity measurements, FLIO imaging shows a characteristic pattern of fluorescence lifetimes over the posterior pole of the retina [3, 4]. The combination of the distribution of fluorescence intensities and lifetimes may reveal additional information about the presence of different retinal fluorophores. Early changes in fluorescence lifetimes may indicate alterations in the metabolic environment.

A topographical map of retinal fluorescence lifetimes in healthy and diseased retina was first described by Schweitzer et al. [5] using the experimental FLIO setup. The first systematic approach with a cohort of 31 healthy subject was published in 2014 by Dysli et al. [6]. They investigated not only the FAF lifetime distribution pattern but also the reproducibility of the measurement, the influence of pupil dilatation, and the impact of age on retinal autofluorescence lifetimes. These findings are in accordance with previous studies using the experimental FLIO, and were later confirmed by other groups [5, 7, 8].

C. Dysli (✉) · M. Dysli
Department of Ophthalmology and Department of Clinical Research, Inselspital, Bern University Hospital, University of Bern, Bern, Switzerland
e-mail: chantal.dysli@insel.ch

© Springer Nature Switzerland AG 2019
M. Zinkernagel, C. Dysli (eds.), *Fluorescence Lifetime Imaging Ophthalmoscopy*, https://doi.org/10.1007/978-3-030-22878-1_8

A characteristic pattern of mean FAF lifetimes can be found in healthy eyes. Figure 6.1 shows this typical distribution pattern of fluorescence lifetimes. In healthy eyes the shortest mean fluorescence lifetimes can be found in the macular center, and are typically depicted in red color within the lifetime distribution color maps. Mean fluorescence lifetimes can be as short as 30 ps at individual pixel spots in the SSC. The central area of the ETDRS grid reveals values between 50 and 200 ps in the SSC, and 100–240 ps in the LSC [6, 8]. Absolute values may vary slightly depending on the device and analyzing technique used in each study (e.g. two- or three-exponential fitting approach and other fitting parameters). This is described in more detail in Chap. 6. Sauer et al. correlated these short lifetime values to the distribution of macular pigment in the central fovea [8]. Outside of the fovea, fluorescence lifetimes are slightly longer. These intermediate decay times may originate from the retinal pigment epithelium [9]. They are colored in orange, yellow and green in FLIO images. Retinal vessels and the optic nerve head with mainly connective tissue components feature significantly longer mean fluorescence lifetimes of approximately 1250 ps in the SSC and 1000 ps in the LSC. These are usually depicted in blue color. The corresponding 2D histogram of the short versus the long decay parameter ($\tau 1$ to $\tau 2$) illustrates clouds of different fluorescence decay times (Fig. 8.1).

With increasing age, the mean fluorescence lifetimes over the entire posterior pole prolong in both spectral channels by about 30 ps per decade in phakic eyes (Fig. 8.2). This increase may be related to the progressive accumulation of lipofuscin and visual cycle byproducts within the retina, resulting in a shift towards fluorophores with longer decay times. Additionally, even though the FLIO device is considered to be a confocal system, significant cataract may influence the mean fluorescence lifetimes either by contribution of autofluorescence from the lens and/ or by scattering of retinal fluorescence. Significant lens opacities lead to blurred FAF intensity images, and to prolongation of fluorescence lifetimes especially within the SSC. However, the LSC is usually only marginally influenced by the lens status, and FAF lifetime changes due to the lens are not as impacting. For clinical studies and quantitative analysis, recording of the lens status is essential, and a comparison of fluorescence lifetime data with subjects of a similar age class and lens status is recommended. When analyzing fluorescence lifetime data from pseudophakic subjects, a prolongation of fluorescence lifetimes with age is only shown for the LSC. This could be due to lipofuscin and its accumulation with age, which is assumed to fluoresce predominantly in the LSC.

FLIO imaging shows high concordance between independent measurements and high reproducibility for quantitative data analysis [7]. Comparison of measurements in dilated and non-dilated pupils reveal slightly longer lifetimes in small pupils, possibly due to a larger amount of lens impact and increased light scattering. Therefore, the current FLIO imaging standard protocol recommends measurements in dilated pupils (see Chap. 6). Nevertheless, FLIO imaging is possible in non-dilated eyes, and may be performed in appropriate situations.

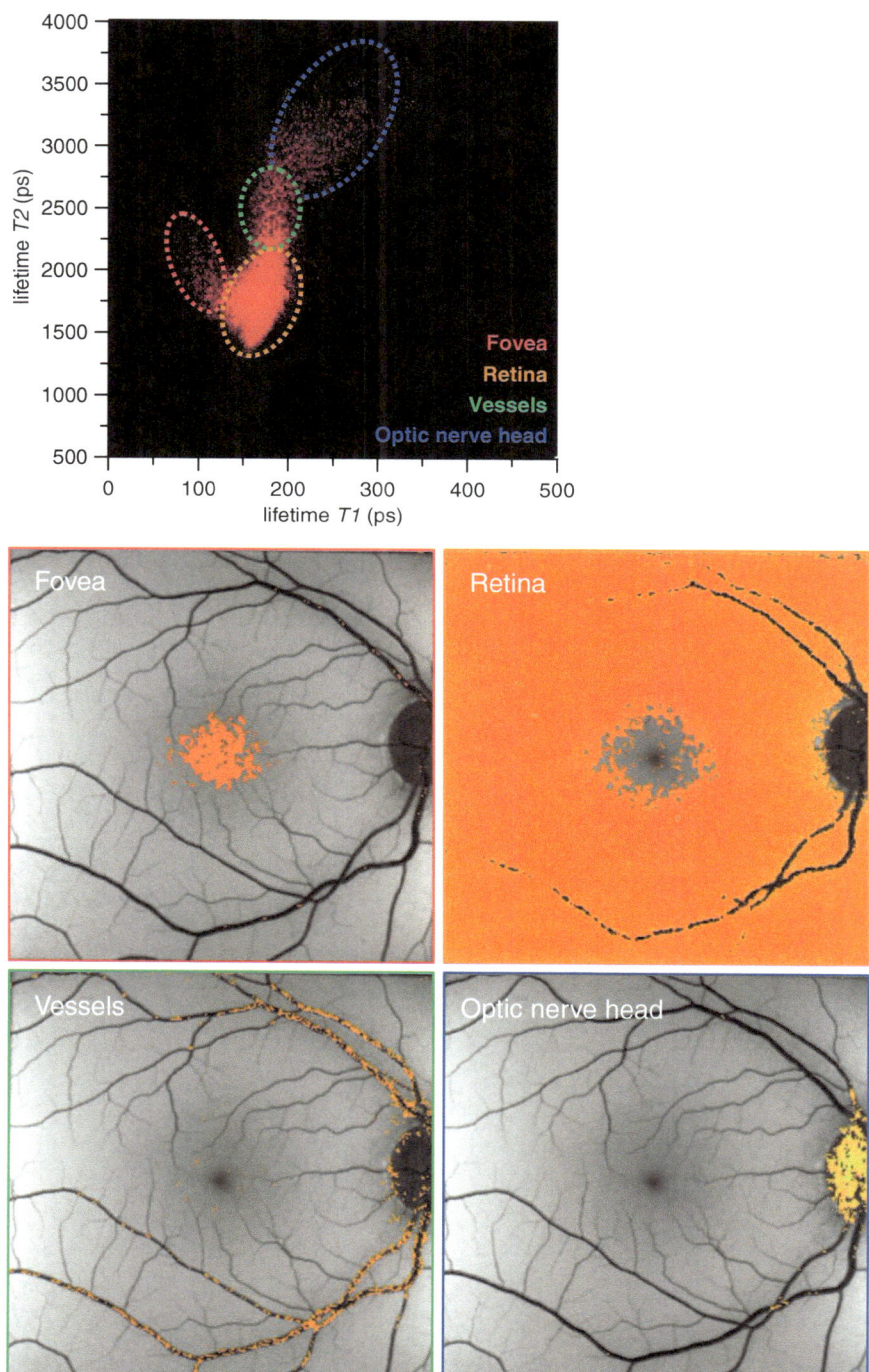

Fig. 8.1 Distribution histogram of the short versus the long decay parameter ($\tau1$ to $\tau2$). In the healthy eye, specific lifetime clouds can be identified for the fovea, the neurosensory retina, retinal vessels, and the optic nerve head as highlighted in separate distribution maps

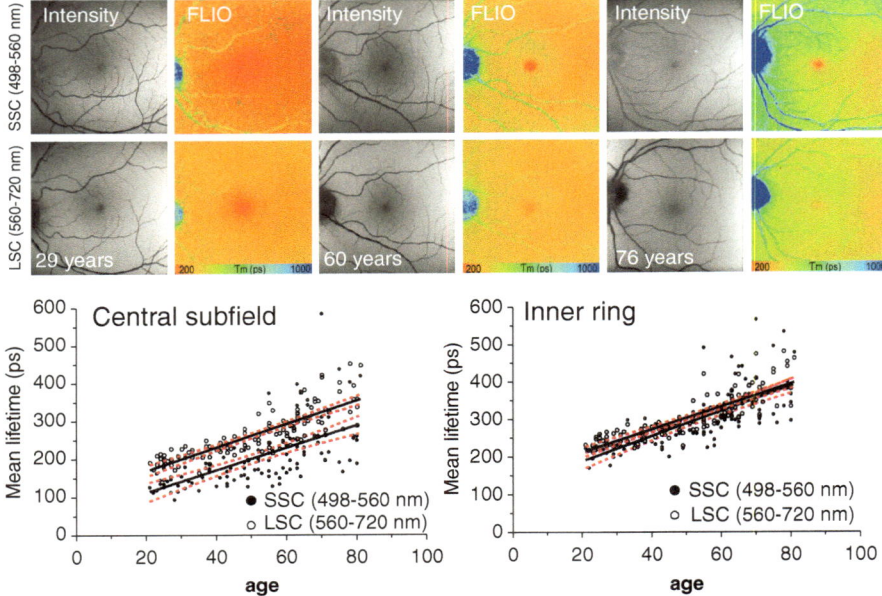

Fig. 8.2 Prolongation of fluorescence lifetimes with age in the retina of healthy eyes. Numerical fluorescence lifetime values are shown in the diagrams below for the central subfield and the inner ETDRS ring for the short (SSC) and the long (LSC) spectral channel with indicated mean and 95% confidence interval

Summary Box

FLIO in healthy retina reveals a characteristic and highly reproducible fluorescence lifetime pattern at the posterior pole. Shortest FAF lifetimes can be found in the center of the macula, probably originating from macular pigment. Intermediate lifetimes are found across the retina, and likely originate from lipofuscin. Longest lifetimes in healthy eyes can be found at retinal vessels and the optic disc, and are assumed to be caused by collagen and elastin. Although FLIO pattern remain the same in the aging population, an overall prolongation of mean fluorescence lifetimes is observed with age. Therefore, age as well as the lens status of the subject should be considered when FLIO data is used for quantitative analysis between different eyes.

References

1. Delori FC, et al. In vivo fluorescence of the ocular fundus exhibits retinal pigment epithelium lipofuscin characteristics. Invest Ophthalmol Vis Sci. 1995;36(3):718–29.
2. Bernstein PS, et al. Lutein, zeaxanthin, and meso-zeaxanthin: the basic and clinical science underlying carotenoid-based nutritional interventions against ocular disease. Prog Retin Eye Res. 2016;50:34–66.

3. Dysli C, et al. Quantitative analysis of fluorescence lifetime measurements of the macula using the fluorescence lifetime imaging ophthalmoscope in healthy subjects. Invest Ophthalmol Vis Sci. 2014;55(4):2106–13.
4. Dysli C, et al. Fluorescence lifetime imaging ophthalmoscopy. Prog Retin Eye Res. 2017;60:120–43.
5. Schweitzer D, et al. Towards metabolic mapping of the human retina. Microsc Res Tech. 2007;70(5):410–9.
6. Dysli C, et al. Quantitative analysis of fluorescence lifetime measurements of the macula using the fluorescence lifetime imaging ophthalmoscope in healthy subjects. Invest Ophthalmol Vis Sci. 2014;55(4):2106–13.
7. Klemm M, et al. Repeatability of autofluorescence lifetime imaging at the human fundus in healthy volunteers. Curr Eye Res. 2013;38(7):793–801.
8. Sauer L, et al. Impact of macular pigment on fundus autofluorescence lifetimes. Invest Ophthalmol Vis Sci. 2015;56(8):4668–79.
9. Sauer L, et al. Monitoring macular pigment changes in macular holes using fluorescence lifetime imaging ophthalmoscopy. Acta Ophthalmol. 2017;95(5):481–92.

Chapter 9
FLIO in Retinal Diseases

Martin Zinkernagel and Chantal Dysli

Retinal imaging is currently the main area of in vivo fluorescence lifetime imaging, and FLIO has been used to obtain information about the lifetimes of endogenous retinal fluorophores in healthy subjects as well as in a variety of retinal diseases [1]. FLIO has been shown to be useful for diagnostics and for understanding of pathophysiologic mechanisms as well as for identifying predictive markers for disease progression and potentially for monitoring of novel therapies. From the current data it has become evident that FLIO provides additional information over other imaging modalities such as FAF or near infrared imaging. In several studies it has been shown that FLIO reliably identifies MP. In another recent publication, supplementation with lutein in selected cases has led to a shortening of fluorescence lifetimes, suggesting that MP can not only be quantified but also monitored in conditions where lutein and zeaxanthin are supplemented [2]. Information about MP can be obtained even in the presence of macular atrophy [3] or other macular pathologies, making FLIO a superior tool to FAF for monitoring of MP in advanced age related macular degeneration [2, 4, 5]. In retinal dystrophies, such as choroideremia and retinitis pigmentosa, FLIO has been proven to be useful to identify remaining photoreceptors even in the absence of atrophy of the retinal pigment epithelium.

When evaluating a FLIO image, deviations from a normal color coded FLIO image should be recognized, and the area of abnormal long or short fluorescence lifetimes can be identified. Abnormal FLIO signal derives either from a change in the composition of fluorophores, but also from a change in the relative amount of these fluorophores. Image acquisition should follow a standardized protocol, and the quality of the image should be checked before interpretation of abnormal FLIO signals, because fluorescence lifetime values are also dependent on the quality of the recorded data. Any opacity of the lens, the cornea or even the vitreous may lead

M. Zinkernagel (✉) · C. Dysli
Department of Ophthalmology and Department of Clinical Research, Inselspital,
Bern University Hospital, University of Bern, Bern, Switzerland
e-mail: martin.zinkernagel@insel.ch

© Springer Nature Switzerland AG 2019
M. Zinkernagel, C. Dysli (eds.), *Fluorescence Lifetime Imaging
Ophthalmoscopy*, https://doi.org/10.1007/978-3-030-22878-1_9

to a general prolongation of the FLIO signal. For the interpretation of FLIO images it is useful to correlate the findings with the FAF images and optical coherence tomography.

Fluorescence lifetime imaging ophthalmoscopy with its simple and noninvasive nature has the potential to find its way into clinical routine. The following chapters will provide an overview of the major findings of retinal imaging with FLIO in various retinal changes and diseases, and summarize the current knowledge and possible applications for the clinical use.

References

1. Dysli C, et al. Fluorescence lifetime imaging ophthalmoscopy. Prog Retin Eye Res. 2017;60:120–43.
2. Sauer L, et al. Fluorescence lifetime imaging ophthalmoscopy (FLIO) of macular pigment. Invest Ophthalmol Vis Sci. 2018;59(7):3094–103.
3. Dysli C, Wolf S, Zinkernagel MS. Autofluorescence lifetimes in geographic atrophy in patients with age-related macular degeneration. Invest Ophthalmol Vis Sci. 2016;57(6):2479–87.
4. Dysli C, et al. Fluorescence lifetimes of Drusen in age-related macular degeneration. Invest Ophthalmol Vis Sci. 2017;58(11):4856–62.
5. Sauer L, et al. Monitoring foveal sparing in geographic atrophy with fluorescence lifetime imaging ophthalmoscopy – a novel approach. Acta Ophthalmol. 2018;96(3):257–66.

Chapter 10
Age-Related Macular Degeneration

Chantal Dysli and Lydia Sauer

Age-related macular degeneration (AMD) is the most common cause for central vision loss in industrial countries within patients over the age of 50 years [1]. In the age group of 50–85 year-olds, a high global prevalence of 8.69% has been reported that increases with increasing age [2]. With ageing global population, the prevalence of AMD is rising [3]. Drusen, extracellular debris deposits located between the basal lamina of the RPE and Bruch membrane's inner collagenous layer, present the hallmark pathological feature of AMD [4, 5]. Complement dysregulation as well as chronic inflammation caused by debris and cellular remnants seem to play a key role in the development of drusen [6–8]. Clinically, AMD can be classified according to the Beckman Initiative for Macular Research Classification Committee as early, intermediate and late AMD [9]. Early AMD is characterized by small drusen (>63 µm but <125 µm) on the posterior pole. Intermediate AMD comprises larger drusen (confluent drusen >125 µm) as well as pigmentary changes. Late AMD is divided in neovascular "exudative" forms with sudden outgrowth of blood vessels (neovascularization), often in combination with leakage of fluid, and "non-exudative" forms with development of geographic atrophy (GA) [10]. Patients typically report decreased central vision, difficulties in dark adaptation, and metamorphopsia. Late forms of AMD may lead to severe vision loss. Intravitreal injections of vascular endothelial growth factor inhibitors (anti-VEGF) may be used for the treatment of the neovascular form of AMD. In contrast, currently, there is no curable treatment method available for

C. Dysli (✉)
Department of Ophthalmology and Department of Clinical Research, Inselspital,
Bern University Hospital, University of Bern, Bern, Switzerland
e-mail: chantal.dysli@insel.ch

L. Sauer
Department of Ophthalmology and Visual Sciences, John A. Moran Eye Center, University of
Utah, Salt Lake City, UT, USA

© Springer Nature Switzerland AG 2019
M. Zinkernagel, C. Dysli (eds.), *Fluorescence Lifetime Imaging
Ophthalmoscopy*, https://doi.org/10.1007/978-3-030-22878-1_10

non-exudative late AMD [11, 12]. In this stage of AMD, delaying of disease progression is currently the primary aim. Development and extensions of geographic atrophy is inevitable and can lead to complete central visual loss [13, 14]. Atrophic areas may initially spare the fovea, relatively normal central visual acuity may be presurved. Currently, several clinical trials are in progress for AMD with GA. Thereby, targeting the complement cascade seems to be one of the most promising approaches [15]. As retinal carotenoid levels seem to be altered in AMD, carotenoid supplementation appears as the only option to slow disease progression in AMD, and a benefit has been reported in clinical studies for intermediate forms of AMD [16–20]. Furthermore, a connection between visual function and the amount of retinal carotenoids has been described for early stages of AMD [21]. Early detection of AMD may therefore provide the opportunity for earlier intervention with carotenoid supplementation, and may potentially slow disease progression and thereby delay or prevent central vision loss.

The exact pathogenesis of AMD is still under investigation. Oxidative stress and subsequent retinal damage seem to play a leading role in disease development. Risk factors for AMD include a genetic profile with gene variations in the complement factor H (CFH) location on chromosome 1 as well as the ARMS 2 locus on chromosome 10 [22, 23]. Additional risk factors reported include advanced age, smoking, and extensive sunlight exposure [4, 24–26].

The most important imaging modality for AMD is currently the optical coherence tomography (OCT), which may detect drusen and sub- or intraretinal fluid, and is also capable to provide comparable follow up images. Further imaging modalities include OCT angiography for neovascular membranes, FA for assessment of neovascular leakage, and FAF intensity imaging as well as fundus photography for visualization of drusenoid deposits as well as geographic atrophy. FAF intensity images provide high contrast due to hypoautofluorescence in absence of the retinal pigment epithelium (RPE).

FLIO emerges as a useful imaging modality in AMD. FLIO measurements are very sensitive and may detect small changes at the human fundus. Through this technology, it may be possible to gain further insights into the pathogenesis of AMD.

FLIO in Early Disease Stages of AMD

It appears to be crucial to detect AMD as early as possible so that the limited therapeutical options can be fully exploited. Metabolic changes may occur in early AMD stages even before they are detectable with conventional imaging techniques [27]. Such metabolic alterations, however, may be detectable with the highly sensitive FLIO modality. Different studies observed changes in fluorescence lifetimes in very early stages of AMD. Patterns of FAF lifetimes in early and intermediate forms of the disease were found to show slightly prolonged fluorescence lifetimes compared to age matches healthy controls [28].

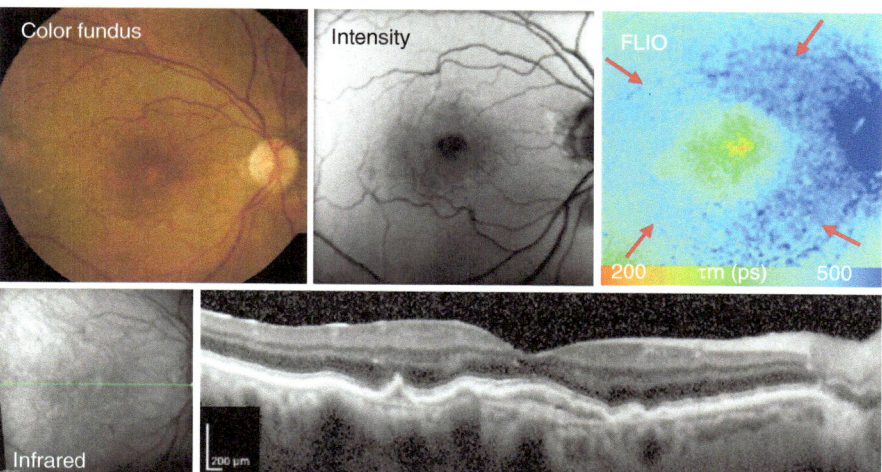

Fig. 10.1 Ring-shaped pattern of prolonged fluorescence lifetimes in a patient with intermediate age-related macular degeneration

Sauer et al. further exploited these changes and reported a ring shaped pattern of prolonged fluorescence lifetimes around the macula, visible in all subjects with AMD [29]. Figure 10.1 shows the typical pattern in an individual with intermediate AMD. It is visible in very early stages of the disease, and also in patients with other retinal diseases in addition to AMD. This observation was especially found with adjusted FLIO color scales ranging from 300 to 500 ps. Other studies showed FLIO images on a scale of 200–1000 ps, a range in which the pattern is less obvious. As this FAF lifetime patterns in AMD can be found already in early stages of AMD it may carry important information towards disease prediction.

As FLIO is a non-invasive in vivo imaging technique, it is difficult to speculate for the cause of the prolongation of lifetimes, and an allocation to individual retinal layers. The observed pattern appears strongest within the LSC, which suggests that the affected fluorophores show their maximal fluorescence emission at wavelengths between 560 and 720 nm. Prolonged fluorescence lifetimes in the LSC may be an early sign for accumulation of bis-retinoids in the RPE. Bis-retinoids feature intrinsic fluorescence, and may also block or diminish the relatively short autofluorescence lifetimes from other fluorophores within the RPE [27, 30]. Early studies on fluorescence lifetimes showed that the LSC is predominantly influenced by lipofuscin [31]. This shift to longer mean FAF lifetimes could indicate that lipofuscin composition may change in early AMD, even before drusen can be found at the fundus. Alterations in lipofuscin content or character may interfere with the healthy retina and cause the disease development of AMD. Sub-RPE basal linear and basal laminar deposits may also cause prolongation of FAF lifetimes, which could reflect an increase of connective tissue due to remodeling processes within the RPE-photoreceptor band [28, 30]. Poorly characterized debris in the sub-retinal space in AMD, that correlated with reticular pseudodrusen, were also reported as well as a spatial correlation of these subretinal deposits with damage to the rod photoreceptors [32, 33].

However, the observation of prolonged lifetimes was also present in 36% of presumably healthy subjects that were age-matched to the AMD patients. These subjects currently feature no symptoms or signs for AMD, but may be of high risk to develop AMD, as most of these individuals, featured positive family history and/or a genetic profile with a very high risk for development of AMD. Longitudinal studies will be necessary to follow up on this observation.

FLIO in AMD with Drusen

Retinal drusen in AMD were recently investigated with FLIO in two different studies [28]. A study by Dysli et al. divided drusen into soft drusen and reticular pseudodrusen based on their appearance on OCT and infrared images. Dysli and colleagues found a general prolongation of FAF lifetimes over the fundus but variable FAF lifetime characteristics in drusen [28]. Reticular pseudodrusen and soft drusen in early and intermediate AMD did not feature distinct fluorescence lifetimes, and were generally not discernible from the surrounding retina. However, subretinal deposits with hyperreflective material in OCT may feature shorter fluorescence lifetimes, whereas intraretinal hyperreflective foci were identified with prolonged fluorescence lifetimes (Fig. 10.2). Similarly, the second study also found that distinguishing drusen using FLIO can be difficult. Although drusen often featured prolonged mean FAF lifetimes, individual drusen may not show a high contrast in FLIO images. Clinical fundus examination and other imaging modalities, such as OCT, FAF intensity imaging, and fundus photography, may be better tools to identify drusen in AMD. However, the presence of short FAF decay times can be aligned to serous pigment epithelial detachment (PED), fibrovascular PED, hemorrhage, subretinal fluid, and choroidal neovascular membrane (unpublished data). This may be useful in the future to differentiate PEDs, hemorrhage and subretinal fluid in AMD.

FLIO in Geographic Atrophy

FLIO measurement in geographic atrophy revealed distinct fluorescence lifetime patterns even in areas of very low autofluorescence intensity [34]. Fluorescence lifetimes in GA are mainly prolonged compared with the surrounding retina (Fig. 10.3). However, towards the macular center, short fluorescence lifetimes may be present which are assumed to origin from remaining macular pigment, and may

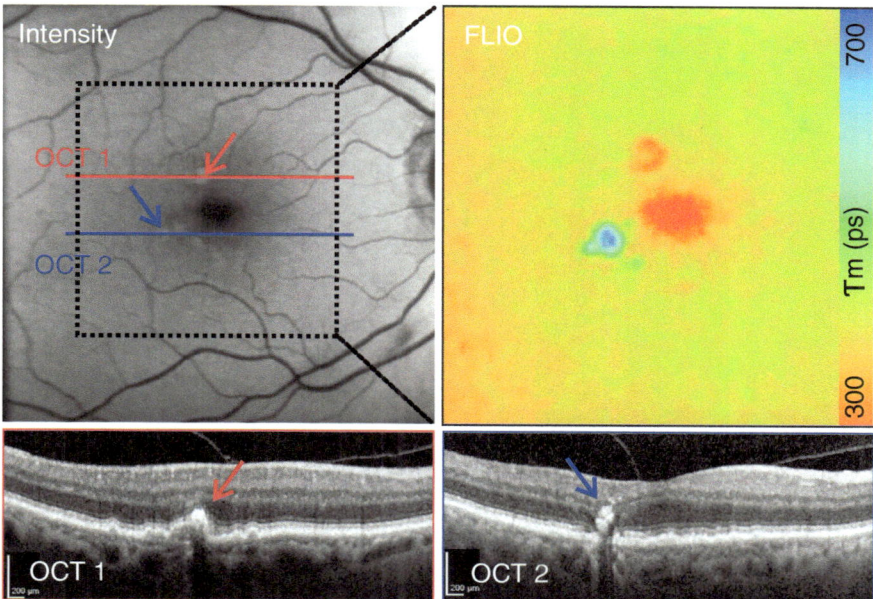

Fig. 10.2 Mean fluorescence lifetimes in drusenoid deposits in a patient with intermediate AMD. Hyperreflective material on the level of the RPE may feature short fluorescence lifetimes, whereas intraretinal hyperreflective foci feature long fluorescence lifetimes. Corresponding OCTs are shown below

indicate persistent outer retinal layers such as the outer nuclear and outer plexiform layer. Additionally, FLIO provides a good contrast in subjects with GA and foveal sparing where the intact retina is clearly identifiable with short fluorescence lifetimes. In detailed analysis including 2D graphs or phasor plots, borders of GA can easily be identified and highlighted. These tools may also provide a good parameter for quantification and follow up examination.

Summary Box
The use of FLIO may facilitate early detection of potential subjects at risk to develop AMD. In early and intermediate AMD stages, fluorescence lifetimes provide additional information about the composition and localization of sub- and intraretinal hyperdense deposits. In late dry AMD, FLIO enables the analysis of hypofluorescent areas, and differentiation of retinal layer structure as well as distribution of macular pigment within and at the border of GA.

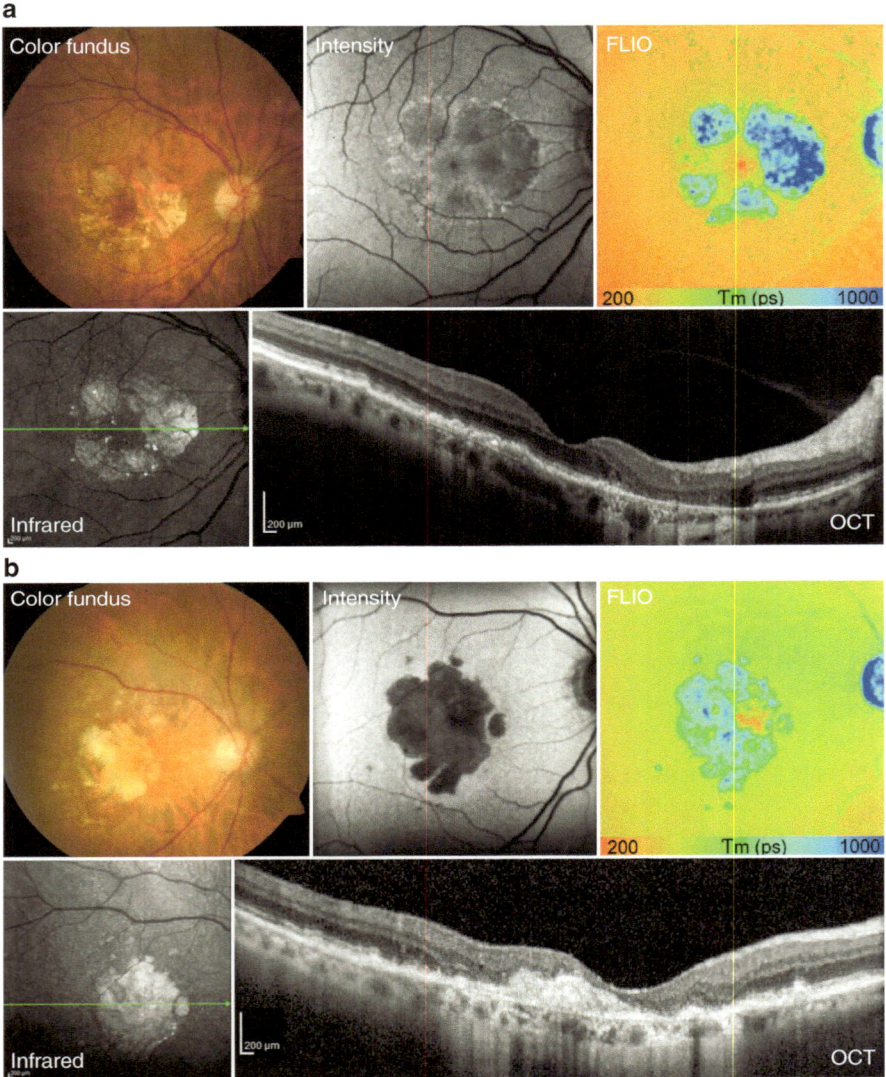

Fig. 10.3 FLIO in two patients with AMD and geographic atrophy (GA). Patient (**a**) features foveal sparing whereas in patient (**b**) GA also involves the foveal center. However, short fluorescence lifetimes are still preserved, probably due to remaining macular pigment

References

1. Jager RD, Mieler WF, Miller JW. Age-related macular degeneration. N Engl J Med. 2008;358(24):2606–17.
2. Wong WL, et al. Global prevalence of age-related macular degeneration and disease burden projection for 2020 and 2040: a systematic review and meta-analysis. Lancet Glob Health. 2014;2(2):e106–16.
3. Friedman DS, et al. Prevalence of age-related macular degeneration in the United States. Arch Ophthalmol. 2004;122(4):564–72.
4. Gehrs KM, et al. Age-related macular degeneration--emerging pathogenetic and therapeutic concepts. Ann Med. 2006;38(7):450–71.
5. Hageman GS, Mullins RF. Molecular composition of drusen as related to substructural phenotype. Mol Vis. 1999;5:28.
6. Hageman GS, et al. An integrated hypothesis that considers drusen as biomarkers of immune-mediated processes at the RPE-Bruch's membrane interface in aging and age-related macular degeneration. Prog Retin Eye Res. 2001;20(6):705–32.
7. Anderson DH, et al. The pivotal role of the complement system in aging and age-related macular degeneration: hypothesis re-visited. Prog Retin Eye Res. 2010;29(2):95–112.
8. Anderson DH, et al. A role for local inflammation in the formation of drusen in the aging eye. Am J Ophthalmol. 2002;134(3):411–31.
9. Ferris FL 3rd, et al. Clinical classification of age-related macular degeneration. Ophthalmology. 2013;120(4):844–51.
10. Sunness JS. The natural history of geographic atrophy, the advanced atrophic form of age-related macular degeneration. Mol Vis. 1999;5:24–35.
11. Rosenfeld PJ, et al. Ranibizumab for neovascular age-related macular degeneration. N Engl J Med. 2006;355(14):1419–31.
12. Chaikitmongkol V, Tadarati M, Bressler NM. Recent approaches to evaluating and monitoring geographic atrophy. Curr Opin Ophthalmol. 2016;27(3):217–23.
13. Maguire P, Vine AK. Geographic atrophy of the retinal pigment epithelium. Am J Ophthalmol. 1986;102(5):621–5.
14. Holz FG, et al. Geographic atrophy: clinical features and potential therapeutic approaches. Ophthalmology. 2014;121(5):1079–91.
15. Boyer DS, et al. The pathophysiology of geographic atrophy secondary to age-related macular degeneration and the complement pathway as a therapeutic target. Retina. 2017;37(5):819–35.
16. Kaya S, et al. Comparison of macular pigment in patients with age-related macular degeneration and healthy control subjects - a study using spectral fundus reflectance. Acta Ophthalmol. 2012;90(5):e399–403.
17. Gorusupudi A, Nelson K, Bernstein PS. The age-related eye disease 2 study: micronutrients in the treatment of macular degeneration. Adv Nutr. 2017;8(1):40–53.
18. Bernstein PS, et al. Lutein, zeaxanthin, and meso-zeaxanthin: the basic and clinical science underlying carotenoid-based nutritional interventions against ocular disease. Prog Retin Eye Res. 2016;50:34–66.
19. Age-Related Eye Disease Study 2 Research Group, et al. Secondary analyses of the effects of lutein/zeaxanthin on age-related macular degeneration progression: AREDS2 report no. 3. JAMA Ophthalmol. 2014;132(2):142–9.
20. Lima VC, Rosen RB, Farah M. Macular pigment in retinal health and disease. Int J Retina Vitreous. 2016;2:19.
21. Akuffo KO, et al. Relationship between macular pigment and visual function in subjects with early age-related macular degeneration. Br J Ophthalmol. 2017;101(2):190–7. https://doi.org/10.1136/bjophthalmol-2016-308418. Epub 2016 Apr 18. PMID: 27091854.
22. Keenan TD, et al. Assessment of proteins associated with complement activation and inflammation in maculae of human donors homozygous risk at chromosome 1 CFH-to-F13B. Invest Ophthalmol Vis Sci. 2015;56(8):4870–9.

23. Hannan JP, et al. Mapping the complement factor H-related protein 1 (CFHR1):C3b/C3d interactions. PLoS One. 2016;11(11):e0166200.
24. Hammond CJ, et al. Genetic influence on early age-related maculopathy: a twin study. Ophthalmology. 2002;109(4):730–6.
25. Seddon JM, Ajani UA, Mitchell BD. Familial aggregation of age-related maculopathy. Am J Ophthalmol. 1997;123(2):199–206.
26. Meyers SM, Greene T, Gutman FA. A twin study of age-related macular degeneration. Am J Ophthalmol. 1995;120(6):757–66.
27. Khan KN, et al. Differentiating drusen: Drusen and drusen-like appearances associated with ageing, age-related macular degeneration, inherited eye disease and other pathological processes. Prog Retin Eye Res. 2016;53:70–106.
28. Dysli C, et al. Fluorescence lifetimes of Drusen in age-related macular degeneration. Invest Ophthalmol Vis Sci. 2017;58(11):4856–62.
29. Sauer L, et al. Patterns of fundus autofluorescence lifetimes in eyes of individuals with nonexudative age-related macular degeneration. Invest Ophthalmol Vis Sci. 2018;59(4):AMD65–77.
30. Dysli C, et al. Fluorescence lifetime imaging ophthalmoscopy. Prog Retin Eye Res. 2017;60:120–43.
31. Schweitzer D, et al. Towards metabolic mapping of the human retina. Microsc Res Tech. 2007;70(5):410–9.
32. Curcio CA, et al. Subretinal drusenoid deposits in non-neovascular age-related macular degeneration: morphology, prevalence, topography, and biogenesis model. Retina. 2013;33(2):265–76.
33. Spaide RF, Curcio CA. Drusen characterization with multimodal imaging. Retina. 2010;30(9):1441–54.
34. Dysli C, Wolf S, Zinkernagel MS. Autofluorescence lifetimes in geographic atrophy in patients with age related macular degeneration. Invest Ophthalmol Vis Sci. 2016;57(6):2479–87.

Chapter 11
FLIO in Diabetic Retinopathy

Lydia Sauer and Martin Hammer

Diabetic retinopathy is a micro-vascular complication in diabetes [1]. Micro-vasculopathy and inflammation finally result in neuronal degeneration [2] and a breakdown of the blood – retina barrier (BRB) causing retinopathy and macular edema [3]. As hyperglycemia is a primary event in diabetes, this causes not only an impairment of the vascular endothelium but also a general protein glycation. This formation of advanced glycation end products (AGEs) in the non-enzymatic Maillard reaction of proteins with glucose and other sugar molecules is involved in BRB breakdown [3], in addition to endothelial dysfunction and the protein-kinase C pathway [4]. As protein glycation is a process generally taking place in ageing tissue and predominantly affecting long-living proteins, it is greatly enhanced in diabetes mellitus [5]. It correlates with the level as well as the duration of hyperglycemia [6]. Protein glycation comprises several steps. First, Schiff's bases are formed in a reaction of the Aldehyde- and Ketone groups of sugars with amino groups of the proteins. This is followed by the Amadori rearrangement finally resulting in the AGE. This is further enhanced by highly reactive Carbonyl groups of intermediates such as α-Oxoaldehyde, Glyoxal, and Methylglyoxal [7]. These intermediates are not only generated by the Maillard reaction but also by other pathways such as auto-oxidation of sugars and glycolysis [8]. Oxidative end products, such as Pentosidin and N-Carboxymethyllysin, and non-oxidative AGEs

L. Sauer
Department of Ophthalmology and Visual Sciences, John A. Moran Eye Center, University of Utah, Salt Lake City, UT, USA
e-mail: Lydia.Sauer@hsc.utah.edu

M. Hammer (✉)
University Hospital Jena, Department of Ophthalmology, Jena, Germany
e-mail: Martin.Hammer@med.uni-jena.de

© Springer Nature Switzerland AG 2019
M. Zinkernagel, C. Dysli (eds.), *Fluorescence Lifetime Imaging Ophthalmoscopy*, https://doi.org/10.1007/978-3-030-22878-1_11

(e.g. Hydroimidazolon and Pyrralin) are distinguished [7, 9]. A well-known AGE is the glycated Hemoglobin HbA1C which is used clinically for long-term monitoring of diabetes [5, 10–12].

Upon hyper-glycemia, AGEs accumulate in the lens, the cornea, the vitreous, and the retina of the eye. Thus, AGEs contribute to diabetic retinopathy in different ways. They may damage the vascular endothelium and, subsequently, also affect the pericytes. This leads to a disruption of the BRB and can result in diabetic macular edema, one of the most sight-threatening complications of diabetes [13, 14]. Furthermore, AGEs have procoagulant potential, contributing to capillary occlusion which is typical for diabetic retinopathy [7]. However, also neuronal cells are affected directly [15]. Animal experiments showed AGE deposition in the vascular as well as in the neuronal compartment of the retina [9]. Protein crosslinking, namely the covalent binding of Lysine residues, alters the tertiary structure of proteins and, thus, impairs their function. However, the modified proteins are able to bind to receptors for advanced glycation end products (RAGE) which are expressed by various cell types such as macrophages, monocytes, endothelial cells, glial cells, and neurons. Besides inflammatory reactions [16], this results in the secretion of cytokines, adhesion molecules, and growth factors like vascular endothelial growth factor (VEGF) [13]. VEGF stimulates neovascularization which is the diagnostic criterion for proliferative diabetic retinopathy. Finally, the activation of the RAGE may excert oxidative stress to the cells by generation of reactive oxygen species (ROS) leading to neuronal cells death [17].

AGE show fluorescence, and their concentration in serum was found to increase with the severity of diabetic retinopathy [18]. As increased fundus autofluorescence (FAF) has been found in diabetic macular edema in association with decreased macular sensitivity [19], Schweitzer et al. [20] and Schmidt et al. [21] investigated fluorescence lifetimes in diabetic patients.

Schweitzer et al. [20] compared the fluorescence decay upon a 448 nm excitation for a group of 48 patients suffering from type 2 diabetes without retinopathy to 48 healthy control subjects of same age. They found a general increase in the fundus autofluorescence lifetimes in diabetes. Using a three-exponential fit of the decay and a sophisticated statistical procedure, they revealed a good discrimination of both groups with a sensitivity of 73% and 70% as well as a specificity of 84% and 64% for the two spectral channels (490–560 nm and 560–720 nm) respectively for the mean fluorescence lifetime τ_m. The best discrimination, however, was achieved by the intermediate decay time component τ_2 at 490–560 nm (sensitivity 84%, specificity 76%) which the authors addressed to fluorophores in the retina. They discuss this as a result of reduced protein binding of FAD as well as protein glycation leading to an accumulation of AGEs. In a subgroup analysis, they found a considerably better discrimination in phakic patients and controls as in pseudo-phakic. Thus, they concluded an influence of the lens fluorescence on the measurements at the fundus despite the use of confocal scanning due to the extremely strong fluorescence emission from the lens. As AGE accumulation in the lens is well known [22], this, in part, could account for the lifetimes measured in diabetics.

Schmidt et al. [21] extended this study to patients with diabetic retinopathy. They compared fluorescence lifetimes upon excitation at 473 nm in 34 patients suffering from non-proliferative diabetic retinopathy (NPDR) with 28 age-matched healthy controls. Fluorescence lifetimes were recorded in the macula as well as two concentric annuli given by the standard ETDRS grid from a three-exponential fit of the decays. Consistent with Schweitzer et al., they showed increased lifetimes in the patient group in all investigated retinal fields (Fig. 11.1). This holds for both spectral channels, however was more pronounced at short wavelengths (498–560 nm, $p < 0.002$) than for longer wavelengths (560–720 nm, $p < 0.05$).

A ROC analysis using a logistic regression model resulted in a sensitivity of 90% and a specificity of 71% for the discrimination of NPDR patients. In contrast to Schweitzer et al., Schmidt et al. found the best discrimination for the long-living fluorescence component τ_3 instead of τ_2. This might result from the longer excitation wavelength used. Again, the formation of AGEs in neurons, vascular, and glial cells was discussed as source of the extension of lifetimes. This was corroborated by FLIO measurements at the lenses of the subjects. These showed shorter lifetimes in the patients, again predominantly in the short wavelength channel. As AGE (bovine serum albumin incubated with glucose) showed a decay time of 1.7 ns and an emission maximum of 523 nm [23], its accumulation must increase the physiologically shorter fundus autofluorescence lifetime, but decrease that of the lens which is known to be longer in healthy state. The assumption of AGEs as source of an additional fluorescence from the ocular fundus is corroborated by the finding of a correlation of the abundance of the intermediate and short lifetime component with the HbA1C value of the patients ($p = 0.009$ and $p = 0.016$ for the short and long wavelength channel respectively, Fig. 11.2).

Taken together, these investigations indicated that FLIO has the potential to show protein glycation as well as alterations in coenzymes of the cellular energy metabolism which are associated with diabetes. This might help in elucidating pathways leading to diabetic retinopathy and, thus, provides opportunities for differential diagnostics with the option of individualized therapy.

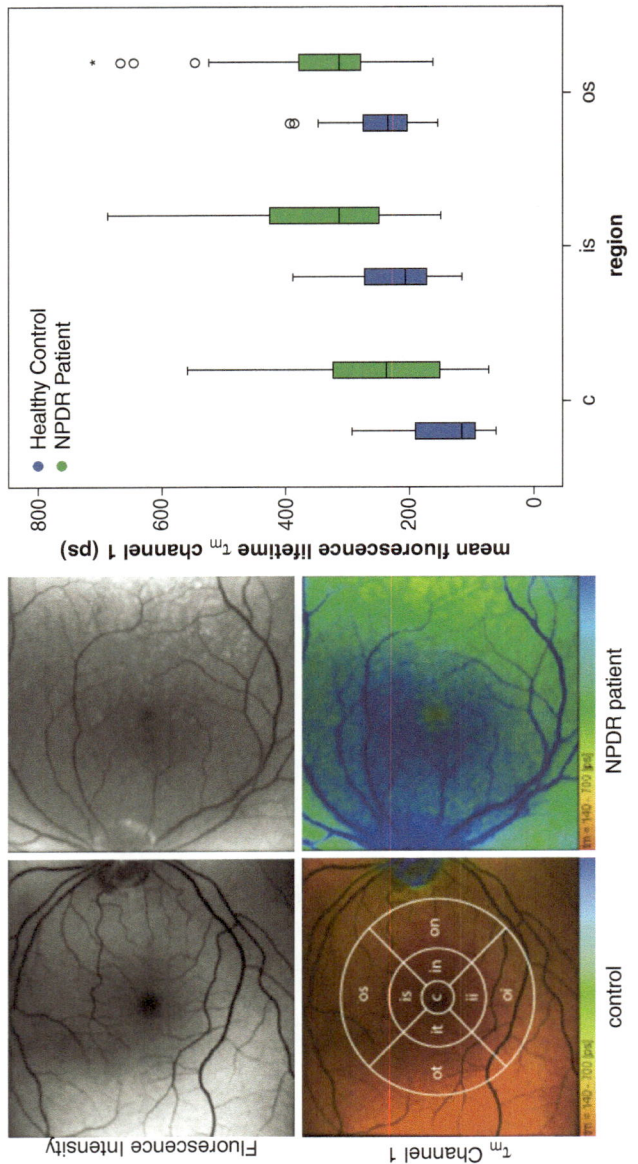

Fig. 11.1 Mean fluorescence lifetimes τ_m in NPDR patients and controls in spectral channel 1 (498–560 nm). **Left:** Autofluorescence intensity (top) and color-coded lifetime images of a control subject (left) and a non-proliferative diabetic retinopathy patient (right). Lower left panel comprises standardized Early Treatment of Diabetic Retinopathy Study (ETDRS)grid (c = central, is = inner superior, in = inner nasal, ii = inner inferior, it = inner temporal, os = outer superior, on = outer nasal, oi = outer inferior, ot = outer temporal). **Right:** Box plots of τ_m in different regions of the retina (c = central, is = inner superior, os = outer superior)

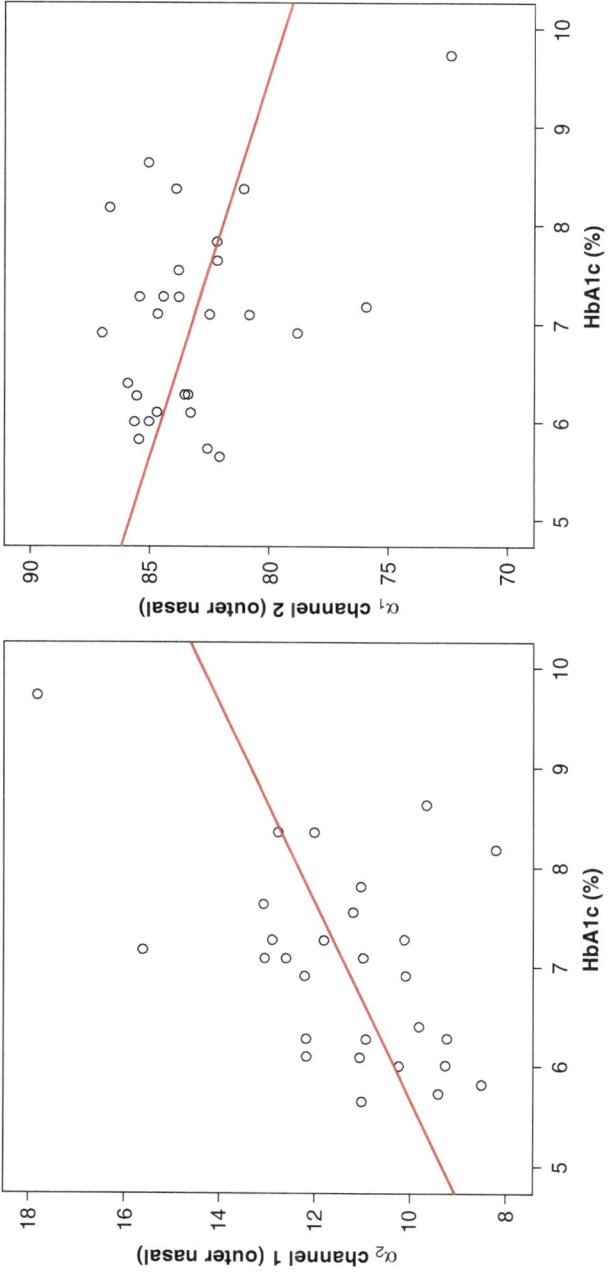

Fig. 11.2 Correlation of α2 in channel 1 and α1 in channel 2 in the outer nasal area, respectively, with HbA1c in the patient group

References

1. Cogan DG, Kuwabara T. Capillary shunts in the pathogenesis of diabetic retinopathy. Diabetes. 1963;12:293–300.
2. Barber AJ, Gardner TW, Abcouwer SF. The significance of vascular and neural apoptosis to the pathology of diabetic retinopathy. Invest Ophthalmol Vis Sci. 2011;52:1156–63.
3. Zhang XY, Zeng H, Bao S, Wang NL, Gillies MC. Diabetic macular edema: new concepts in patho-physiology and treatment. Cell Biosci. 2014;4:27.
4. Yu DY, Cringle SJ, Su EN, Yu PK, Jerums G, Cooper ME. Pathogenesis and intervention strategies in diabetic retinopathy. Clin Exp Ophthalmol. 2001;29:164–6.
5. Tarr JM, Kaul K, Chopra M, Kohner EM, Chibber R. Pathophysiology of diabetic retinopathy. ISRN Ophthalmol. 2013;2013:343560.
6. Vlassara H, Palace MR. Diabetes and advanced glycation endproducts. J Intern Med. 2002;251:87–101.
7. Singh R, Barden A, Mori T, Beilin L. Advanced glycation end-products: a review. Diabetologia. 2001;44:129–46.
8. Behl T, Kaur I, Kotwani A. Implication of oxidative stress in progression of diabetic retinopathy. Surv Ophthalmol. 2016;61(2):187–96.
9. Hammes HP, Alt A, Niwa T, et al. Differential accumulation of advanced glycation end products in the course of diabetic retinopathy. Diabetologia. 1999;42:728–36.
10. Glenn JV, Stitt AW. The role of advanced glycation end products in retinal ageing and disease. Biochim Biophys Acta. 2009;1790:1109–16.
11. Stitt AW. Advanced glycation: an important pathological event in diabetic and age related ocular disease. Br J Ophthalmol. 2001;85:746–53.
12. Stitt AW. AGEs and diabetic retinopathy. Invest Ophthalmol Vis Sci. 2010;51:4867–74.
13. Kandarakis SA, Piperi C, Topouzis F, Papavassiliou AG. Emerging role of advanced glycation-end products (AGEs) in the pathobiology of eye diseases. Prog Retin Eye Res. 2014;42:85–102.
14. Calvo P, Abadia B, Ferreras A, Ruiz-Moreno O, Verdes G, Pablo LE. Diabetic macular edema: options for adjunct therapy. Drugs. 2015;75:1461–9.
15. Eisma JH, Dulle JE, Fort PE. Current knowledge on diabetic retinopathy from human donor tissues. World J Diabetes. 2015;6:312–20.
16. Hernandez C, Dal Monte M, Simo R, Casini G. Neuroprotection as a therapeutic target for diabetic retinopathy. J Diabetes Res. 2016;2016:9508541.
17. Yamagishi S, Maeda S, Matsui T, Ueda S, Fukami K, Okuda S. Role of advanced glycation end products (AGEs) and oxidative stress in vascular complications in diabetes. Biochim Biophys Acta. 2012;1820:663–71.
18. de la Maza MP, Garrido F, Escalante N, et al. Fluorescent advanced glycation end-products (ages) detected by spectro-photofluorimetry, as a screening tool to detect diabetic microvascular complications. J Diabetes Mellitus. 2012;2:221–6.
19. Vujosevic S, Casciano M, Pilotto E, Boccassini B, Varano M, Midena E. Diabetic macular edema: fundus autofluorescence and functional correlations. Invest Ophthalmol Vis Sci. 2011;52:442–8.
20. Schweitzer D, Deutsch L, Klemm M, et al. Fluorescence lifetime imaging ophthalmoscopy in type 2 diabetic patients who have no signs of diabetic retinopathy. J Biomed Opt. 2015;20:61106.
21. Schmidt J, Peters S, Sauer L, et al. Fundus autofluorescence lifetimes are increased in nonproliferative diabetic retinopathy. Acta Ophthalmol. 2017;95(1):33–40.
22. Araki N, Ueno N, Chakrabarti B, Morino Y, Horiuchi S. Immunochemical evidence for the presence of advanced glycation end products in human lens proteins and its positive correlation with aging. J Biol Chem. 1992;267:10211–4.
23. Schweitzer D, Schenke S, Hammer M, et al. Towards metabolic mapping of the human retina. Microsc Res Tech. 2007;70:410–9.

Chapter 12
Retinal Artery Occlusion

Martin Zinkernagel and Chantal Dysli

Retinal artery occlusion is one of the most frequently encountered reasons for acute vision loss in the elderly. In addition to the visual impairment, it is associated with high mortality due to other comorbidities such as atherosclerosis. Occlusion of the retinal arteries lead to ischemia and edema of the inner retina in the acute phase. Over time, the edema resolves and a generalized atrophy of the inner retinal layers remains. Several studies have identified distinctive features of retinal ischemia in retinal artery occlusion in OCT. Initially, a hyperreflectivity with loss of the organized retinal layers can be observed within the inner retina [1]. In the later stages, a generalized atrophy of the inner layers, which are supplied by retinal arteries, can be seen in OCT. The use of fundus autofluorescence to investigate ischemic areas after retinal artery occlusion has only received very little attention so far [2]. However, there are several studies showing that fundus autofluorescence may help to detect and differentiate retinal emboli. Especially atherosclerotic emboli, which consist for a large part of lipofuscin, are well visible as hyperautofluorescent lesions in fundus autofluorescence [3]. Fluorescein angiography in retinal artery occlusion is seldom useful because perfusion often reverts to normal even after complete retinal artery occlusion [4]. Furthermore, perfusion insufficiency does not always mirror tissue ischemia because retinal cells have different oxygen consumption rates and may have different susceptibility to hypoxia [5].

A sensitive noninvasive imaging technique capable of detecting early tissue alterations consistent with retinal hypoxia or even ischemia could potentially improve therapeutic options and may influence outcomes.

Although still in the early stage of development for clinical applications, fluorescent techniques based on the detection of endogenous fluorophores such as FLIO could provide noninvasive methods for diagnosis of retinal ischemic disease [6]. The time-domain characteristics of fluorescence, rather than its intensity, have been

M. Zinkernagel (✉) · C. Dysli
Department of Ophthalmology and Department of Clinical Research, Inselspital, Bern University Hospital, University of Bern, Bern, Switzerland
e-mail: martin.zinkernagel@insel.ch

© Springer Nature Switzerland AG 2019
M. Zinkernagel, C. Dysli (eds.), *Fluorescence Lifetime Imaging Ophthalmoscopy*, https://doi.org/10.1007/978-3-030-22878-1_12

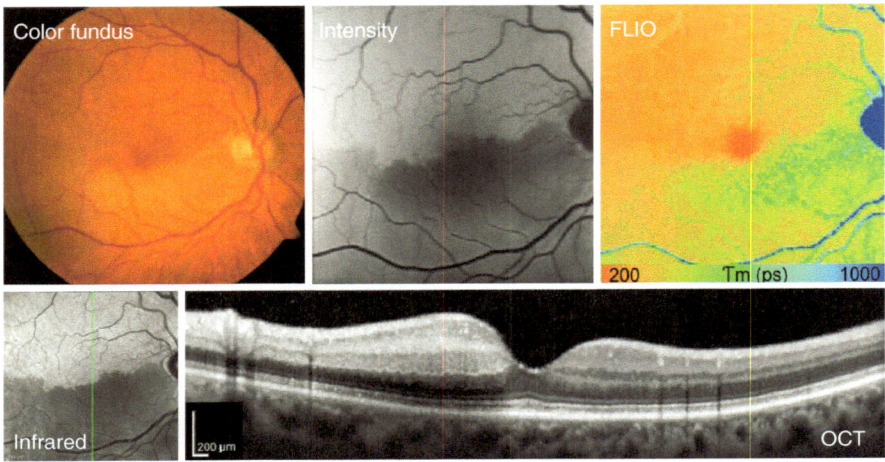

Fig. 12.1 Multimodal imaging in a patient with inferior branch retinal artery occlusion with swelling of the inner retinal layers and prolongation of the corresponding fluorescence lifetimes

exploited for imaging applications in microscopy [7] and FLIM has been shown to be sufficiently sensitive to indicate ischemic kidney ex vivo in an animal model [8]. In vivo imaging of ischemia is often difficult to perform due to poor tissue penetration, especially when using fluorophores with short emission wavelengths.

Because of the optical properties of the eye, issues with tissue penetration are virtually absent. Fluorescence lifetime imaging in patients with retinal artery occlusion revealed a significant prolongation of fluorescence lifetimes in the acute phase when compared to an age matched healthy cohort (Fig. 12.1) [6]. The prolongation was found for both spectral channels and lasting up to 3 months after the incidence. Thereafter, fluorescence lifetime values returned to normal levels despite marked atrophy of the inner retina in optical coherence tomography. Several factors may explain the prolongation of fluorescence lifetimes in acute retinal artery occlusion. First of all, retinal ganglion layer swelling may block the underlying fluorescence and therefore could distort fluorescence lifetime signals from the outer retinal layers and the retinal pigment epithelium. Another explanation could be that changes in the cellular metabolism caused by retinal hypoxia lead to prolongation in fluorescence lifetimes. In this context, changes in the equilibrium of redox pairs NAD+/NADH (oxidized and reduced nicotinamide adenine dinucleotide) and FAD/FADH$_2$ (oxidized and reduced flavin adenine dinucleotide) could potentially explain these observations. After the acute phase, fluorescence lifetimes return to levels found in healthy fellow eyes. Although at first sight a negative finding, it suggest that the inner retina contributes relatively little to fluorescence lifetimes seen with FLIO.

Summary Box

In conclusion, FLIO can be used to identify ischemic retinal changes especially in the acute phase of retinal artery occlusion in the area of edema of the inner retinal layers. Fluorescence lifetimes return to levels found in healthy eyes after atrophy of the inner retinal layers has developed.

References

1. Chu YK, et al. In vivo detection of acute ischemic damages in retinal arterial occlusion with optical coherence tomography: a "prominent middle limiting membrane sign". Retina. 2013;33(10):2110–7.
2. Mathew R, Papavasileiou E, Sivaprasad S. Autofluorescence and high-definition optical coherence tomography of retinal artery occlusions. Clin Ophthalmol. 2010;4:1159–63.
3. Bacquet JL, et al. Fundus autofluorescence in retinal artery occlusion: a more precise diagnosis. J Fr Ophtalmol. 2017;40(8):648–53.
4. Hayreh SS. Prevalent misconceptions about acute retinal vascular occlusive disorders. Prog Retin Eye Res. 2005;24(4):493–519.
5. Bek T. Inner retinal ischaemia: current understanding and needs for further investigations. Acta Ophthalmol. 2009;87(4):362–7.
6. Dysli C, Wolf S, Zinkernagel MS. Fluorescence lifetime imaging in retinal artery occlusion. Invest Ophthalmol Vis Sci. 2015;56(5):3329–36. https://doi.org/10.1167/iovs.14-16203.
7. Day RN, Schaufele F. Imaging molecular interactions in living cells. Mol Endocrinol. 2005;19(7):1675–86.
8. Abulrob A, et al. In vivo time domain optical imaging of renal ischemia-reperfusion injury: discrimination based on fluorescence lifetime. Mol Imaging. 2007;6(5):304–14.

Chapter 13
Central Serous Chorioretinopathy

Chantal Dysli and Martin Zinkernagel

Central serous chorioretinopathy (CSCR) is a retinal disorder characterized by idiopathic serous detachment of the neurosensory retina and occasionally associated with focal RPE detachment [1]. CSCR belongs to the pachychoroid disease spectrum according to the marked thickening of the choroidea, [2] and is more frequent in hyperopic eyes. It typically affects middle-aged male subjects (72–88%) between 39 and 51 years, and is thought to be associated with the corticosteroid metabolism. Multiple risk factors are discussed including predisposing genetic profile, systemic factors such as arterial hypertension, psychological and personality profile, sympathetic-parasympathetic imbalance, corticosteroids or other drugs, endocrine changes, sleep disturbance, and others. However, the exact disease mechanism is not yet completely solved and has to be further investigated.

As first symptoms, decreased visual acuity as well as metamorphopsia, micropsia, central scotoma, dyschromatopsia, and reduced contrast sensitivity may be reported. In the majority of cases, the disease is self-limited and resolves over 3–6 months [3]. However, if leakage persists over 4–6 months, CSCR may convert into a chronic form either with continuous subretinal fluid or with chronic recurrence of disease symptoms. In such cases, excessive pigmentary changes, atrophy of the neurosensory retina, and even secondary neovascularization may occur.

Funduscopically, fluid accumulation as well as pigmentary changes might be observed. Additional imaging modalities such as OCT, FAF, fluorescein and indocyanine angiography are useful for confirmation of the diagnosis as well as for follow-up examinations.

Even though CSCR classically manifests primarily in one eye, it is a bilateral disease with minor changes possibly without clinical symptoms on the fellow eye.

C. Dysli (✉) · M. Zinkernagel
Department of Ophthalmology and Department of Clinical Research, Inselspital,
Bern University Hospital, University of Bern, Bern, Switzerland
e-mail: chantal.dysli@insel.ch

© Springer Nature Switzerland AG 2019
M. Zinkernagel, C. Dysli (eds.), *Fluorescence Lifetime Imaging Ophthalmoscopy*, https://doi.org/10.1007/978-3-030-22878-1_13

However, future appearance of clinical symptoms and disease progression in the contralateral eye is still possible.

By now, no causative treatment is available. Generally, avoidance of possible risk factors is advised. In the acute disease stage, primary observation is recommended. Treatment approaches include mineralocorticoid receptor antagonists, photodynamic therapy with verteporfin, focal laser therapy, and intravitreal anti-VEGF if secondary neovascularizations are present.

In a study of Dysli et al., 35 subjects (mean age 46 ± 6 years) with acute or chronic (>6 months) CSCR, and a corresponding age-matched healthy control cohort was investigated using FLIO [4]. Thereby, FLIO was shown to be a very sensitive tool for detection of retinal changes. Even before a circle of hyperautofluorescence is visible in FAF intensity images, FLIO might show disease specific findings. Interestingly, the amount and presence or absence of subretinal fluid at time of measurement did not directly influence the measured fluorescence lifetimes. However, in the acute disease stage, large areas of short fluorescence lifetimes are observed, color coded in red (Fig. 13.1). They were associated with elongated outer photoreceptor segments and might origin from accumulated visual cycle end products due to increased distance to the RPE and interrupted phagocytosis of disc outer segments. Over time, the autofluorescence lifetime values normalize towards values of the unaffected retina. However, in chronic disease stage, secondary retinal changes such as scar formation and retinal atrophy with loss of the RPE and the photoreceptor layer lead towards prolonged FLIO lifetime values (Fig. 13.2). This is comparable to changes observed in geographic atrophy in AMD [5, 6].

Summary Box
FLIO in CSCR features a characteristic distribution pattern with short lifetimes in areas of disease activity, changing to prolonged lifetimes over time, corresponding to secondary remodeling of the retinal layer structure and scar formation. FLIO was shown to be sensitive to early and subtle retinal changes in CSCR, and may provide information about the integrity of the photoreceptors and the RPE.

Fig 13.1

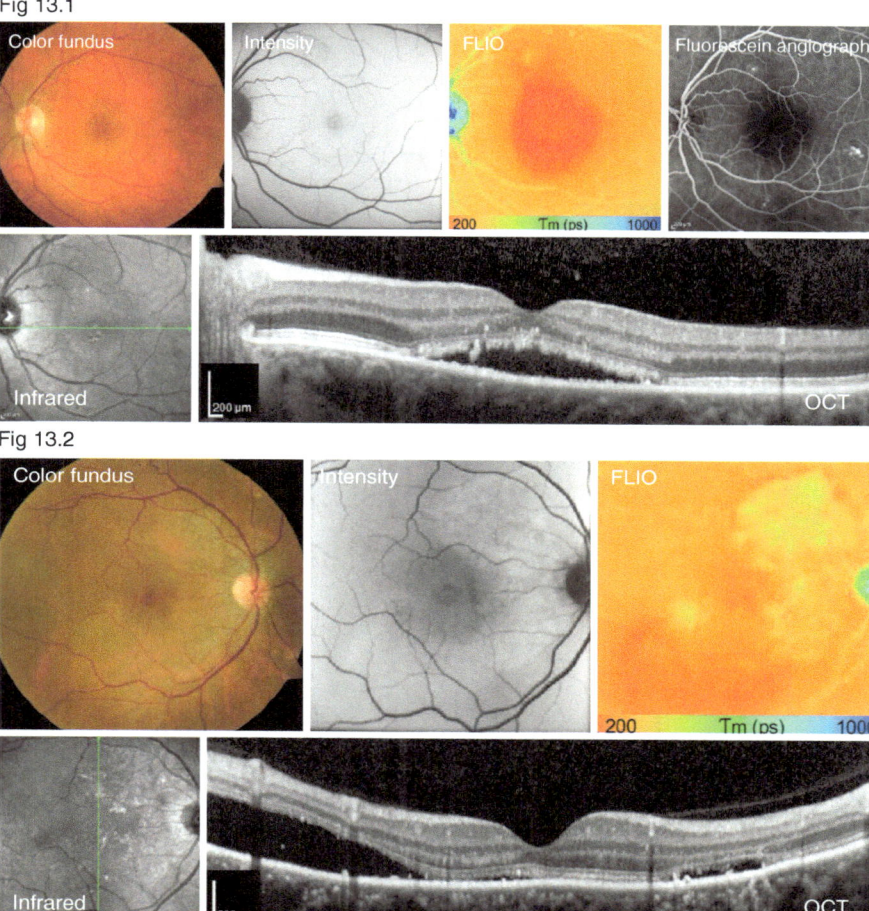

Fig 13.2

Figs. 13.1 and 13.2 Multimodal imaging in patients with central serous chorioretinopathy (CSCR). Subject 13.1 shows an acute disease stage with clearly visible short fluorescence lifetimes in the affected area. The corresponding fluorescence intensity image shows only very subtle changes. Patient 13.2 currently features disease activity inferiorely with short fluorescence lifetimes, and superiorly prolonged lifetimes due to long-term changes in chronic CSCR

References

1. Daruich A, et al. Central serous chorioretinopathy: recent findings and new physiopathology hypothesis. Prog Retin Eye Res. 2015;48:82–118.
2. Akkaya S. Spectrum of pachychoroid diseases. Int Ophthalmol. 2018;38(5):2239–46.
3. Peiretti E, et al. Anti-vascular endothelial growth factor therapy versus photodynamic therapy in the treatment of choroidal neovascularization secondary to central serous chorioretinopathy. Retina. 2018;38(8):1526–32.
4. Dysli C, et al. Fundus autofluorescence lifetimes and central serous chorioretinopathy. Retina. 2017;37:2151.
5. Dysli C, et al. Fluorescence lifetime imaging ophthalmoscopy. Prog Retin Eye Res. 2017;60: 120–43.
6. Dysli C, Wolf S, Zinkernagel MS. Autofluorescence lifetimes in geographic atrophy in patients with age related macular degeneration. Invest Ophthalmol Vis Sci. 2016;57(6):2479–87.

Chapter 14
Macular Telangiectasia Type 2

Lydia Sauer and Paul S. Bernstein

Macular telangiectasia type 2 (MacTel) is an inherited retinal disease that typically starts to affect the vision in patients between 40 and 60 years [1–3]. However, cases of younger subjects have been reported in recent literature, and the prevalence of 0.1% is likely underestimated. Patients may often be misdiagnosed with AMD [4]. In contrast to macular telangiectasia type 1, which usually affects young males unilaterally, MacTel is a bilateral disease, causing metamorphopsia and central vision loss. Problems with reading small print due to letters that seem missing is a common complaint of patients with MacTel [5]. Although the disease typically does not proceed to legal blindness, the vision of most patients becomes significantly disturbed over time.

MacTel was first described by Gass in 1977, where he called it 'bilateral paracentral capillary telangiectasia of unknown cause' [6]. The term 'telangiectasia' refers to dilated and abnormal vessels, that may form in the course of the disease. In 1993, Gass and Bondi published another article, which further classified different types of retinal telangiectasia [7]. MacTel is consistent with the type 2A that was described in this article. Ectatic vessels, especially capillaries in the deeper capillary network as well as dilated venules are typical findings in this disease. However, evidence is rising that telangiectasia of retinal vessels is not the origin but rather a secondary feature of MacTel. Initially, retinal greying and the loss of a foveal reflex are described as typical signs of disease onset. Abnormal low macular pigment is another typical feature of the disease. Although it has been reported that patients may show MP levels similar to healthy levels in early stages of MacTel, patients with advanced disease show severely reduced MP levels [8]. Furthermore, MP shows abnormal distributions at 5–9 degree eccentricity from the fovea [9–13]. If these patients supplement carotenoids, the central MP cannot be re-established. Instead, the ring-like distribution of MP outside of the fovea is enhanced by supplementing Lutein and Zeaxanthin [11, 14, 15]. Based on these findings, a loss

L. Sauer (✉) · P. S. Bernstein
University of Utah, John A. Moran Eye Center, Salt Lake City, UT, USA
e-mail: Lydia.Sauer@hsc.utah.edu

© Springer Nature Switzerland AG 2019
M. Zinkernagel, C. Dysli (eds.), *Fluorescence Lifetime Imaging Ophthalmoscopy*, https://doi.org/10.1007/978-3-030-22878-1_14

of Müller- and photoreceptor cells may be key factors in the development of MacTel [16, 17]. Another early and typical feature of MacTel is a loss of the ellipsoid zone, which seems to be associated with a loss of retinal sensitivity as well as central visual acuity [18, 19]. Almost half of the patients with MacTel also presents with intraretinal crystals, however, the cause of these crystals is still unclear [7, 20–22]. In late stages of disease, retinal pigment clumps can be found. These may be associated with foveal atrophy in late stages of MacTel [7].

It is very typical that retinal changes associated with MacTel occur in a well defined area, the so-called 'MacTel zone'. This zone was described as an oval region centered around the fovea, with eccentricities of 5° vertically and 6° horizontally. Changes associated with MacTel often start in the area temporal to the fovea, which is referred to as the para-foveal area [23]. These alterations may eventually affect the entire MacTel zone. Nevertheless, the temporal para-foveal area appears to be most severely affected [24–26].

Recent literature indicates very promising results towards novel therapeutical approaches for MacTel. In the past, patients were occasionally treated with intra-vitreal injections of steroids or anti-VEGF, but neither of these approaches improved visual outcome [27–29]. However, the effect of ciliary neurotrophic factor is currently being investigated and showed very promising results in early clinical trials [30, 31]. This therapy - an implant that releases ciliary neurotrophic factor - may be very beneficial for patients. With approaching treatment possibilities, it becomes even more important to diagnose MacTel as early as possible.

Despite extensive research on MacTel, the pathogenesis behind the disease remains unknown. Recent studies showed that MacTel likely has a dominant genetic inheritance with reduced penetrance [32–34]. Recently two genetic loci were identified for MacTel. These are related to the glycine/serine pathway, which appears to be disturbed in the disease. However, the actual genes which cause MacTel have not yet been found. Therefore, genetic testing is not available and the diagnose of MacTel remains clinical [32, 35, 36].

Multimodal imaging is the key to diagnose MacTel at early stages. Both, functional as well as structural imaging modalities can be used to investigate patients with signs of MacTel. However, due to the variability of features between patients it is challenging for many ophthalmologists to correctly make the diagnose of MacTel in early disease stages. The unique changes of MacTel can be observed with a variety of different imaging modalities. Color fundus photography may show retinal greying as well as the loss of the foveal reflex. OCT imaging often shows inner-retinal cysts, which commonly present at the temporal para-foveal area. Despite being a very typical finding, retinal cysts may be absent in early disease stages. Later in the course of disease, breaks within the inner-segment/outer-segment (IS/OS) junction can be observed with OCT imaging. Blue-light reflectance imaging of eyes with MacTel shows a bulls-eye lesion, and fundus autofluorescence often presents with a hyperfluorescent foveal region due to the absence of macular pigment [37–40]. The central hyperfluorescence has been associated with poorer central visual acuity [41]. MP imaging typically shows low MP values with occasional abnormal MP enhancement outside of the MacTel zone. Microperimetry testing, seems to be a promising tool for monitoring progression of vision loss in patients with MacTel [42]. Previous studies using microperimetry in MacTel patients revealed that the dysfunction of rods seems to be more severe than that of cones [43]. Historically, fluorescein angiography was seen as the gold-standard to image MacTel. However, fluorescein angiography required the intra-venous injection of fluorescent dye, which shows leakage in the area affected with MacTel. The leakage typically occurs within the temporal para-foveal area. As this imaging technique is invasive, non-invasive imaging should be preferred [7, 37, 44–46].

FLIO is a novel tool on the list of imaging modalities to detect MacTel. FLIO has two clear advantages: (1) it is non-invasive, and (2) it shows a very characteristic pattern in all eyes with MacTel. The first study using FLIO to investigate MacTel was published in 2018 [47]. This initial study included 42 eyes of 21 patients with different stages of MacTel. Another study conducted at a different FLIO center investigated 14 patients and was able to confirm these findings [48]. Further studies of more than 70 patients with MacTel are also confirmative of these initial findings (unpublished data). In all eyes with MacTel, a very typical FLIO pattern was observed: FLIO lifetimes were characteristically prolonged in the MacTel zone, predominantly within the SSC. Figure 14.1 shows this pattern.

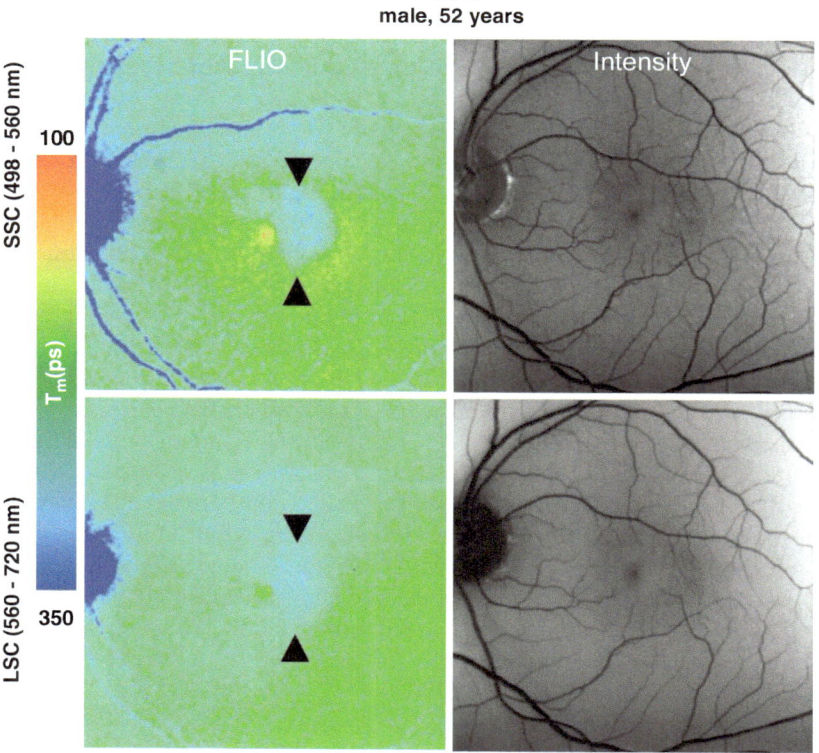

Fig. 14.1 FLIO images and fluorescence intensity images of a patient with MacTel. Arrowheads point towards the typical MacTel pattern

Typically, MacTel eyes showed a crescent-shaped prolongation of FLIO lifetimes at the temporal side of the fovea. These prolonged FLIO lifetimes were described to surround the fovea and include the entire MacTel area in more advanced MacTel stages. It was also shown that FLIO lifetimes prolong with disease progression over time. FLIO shows this characteristic pattern of MacTel with high contrast.

FLIO is based on fluorescence imaging, recording not only the intensity but also the lifetime of the fluorescence. Therefore, it provides additional information to fundus autofluorescence intensity imaging. Molecules with a weak fluorescence, such as macular pigment, can be detected with FLIO [49, 50]. It therefore is not surprising that FLIO highlights the altered macular pigment, especially if MP accumulates around the MacTel zone. The central area of the fovea, on the other hand, shows longer fluorescence lifetimes due to reduced macular pigment levels. But the detection of FLIO lifetimes not only shows altered macular pigment but may also show subtle biochemical changes within the retinal molecules, potentially even before damage to the retina occurs. It was shown that MacTel crystals correspond to prolonged FLIO lifetimes, which confirms that these crystals are likely not macular pigment [47]. However, the exact composition of these crystals remains unclear. In addition to the FLIO findings in MacTel, characteristic patterns different from the MacTel pattern have been reported for many other retinal diseases [51–58]. Therefore, FLIO appears to be helpful to distinguish MacTel from other retinal diseases that may clinically show similar features, for example AMD.

FLIO lifetimes in MacTel were compared to groups of age-matched healthy eyes. In contrast to these healthy eyes, the temporal para-foveal area T1 (inner temporal area of a standardized ETDRS grid) shows significantly prolonged FLIO lifetimes ($p < 0.001$). This was compared to a reference area T2 (outer temporal area of a standardized ETDRS grid), which did not show differences between these groups. It was therefore suggested that a ratio of T2/T1 of less than 0.9 in combination with typical disease features may be indicative of MacTel [47].

Of special interest are changes within FLIO lifetimes that may be detectable in family members of MacTel patients that do not show clinical evidence of the disease. FLIO has already been reported for one family where both parents of a MacTel patient were clinically healthy [47]. If an autosomal dominant inheritance is assumed for MacTel, at least one of the parents would have to carry a putative MacTel gene. Although the clinical exam as well as conventional imaging was normal for the mother of this patient, she showed prolonged FLIO lifetimes that may be related to subtle alterations due to MacTel. Figure 14.2 represents these findings. Further studies investigated larger numbers of family members (unpublished data). Here, prolonged FLIO lifetimes were found in more family members. Some of these family members are highly suspicious for early MacTel. Figure 14.3 highlights these findings.

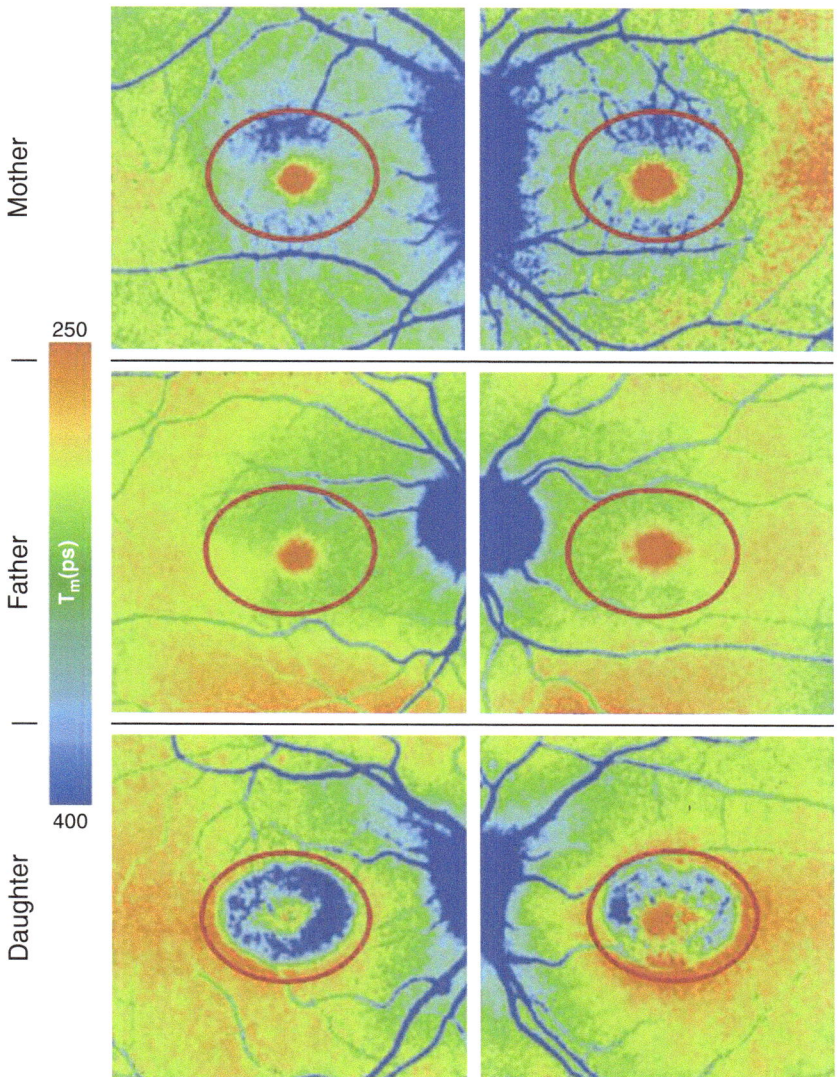

Fig. 14.2 Fluorescein angiography as well as Mean FLIO lifetimes (short spectral channel) of a family with macular telangiectasia type 2 (MacTel). The mother shows prolonged FLIO lifetimes that could be indicative of MacTel as well as genetic predisposition

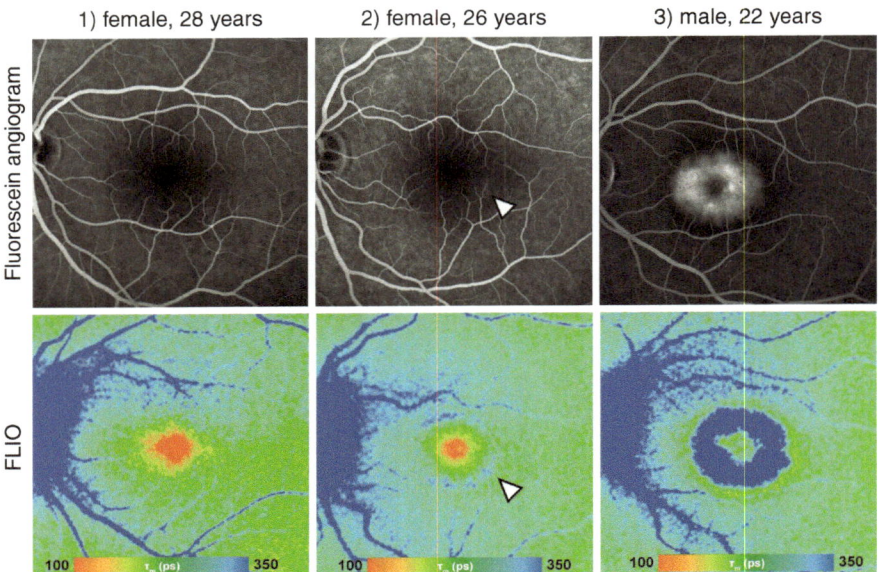

Fig. 14.3 Fluorescein angiography as well as Mean FLIO lifetimes (short spectral channel) of a family with macular telangiectasia type 2 (MacTel). Whereas subject 3) has been diagnosed with MacTel, his two sisters (1/2) are clinically unremarkable. One of his sisters (2) shows FLIO lifetimes indicative of MacTel as well as genetic predisposition

Summary Box

To conclude, FLIO highlights retinal changes in MacTel. A very typical pattern of prolonged FLIO lifetimes can be found in all patients with MacTel. Therefore, not only the reduced central and abnormally increased peripheral accumulation of macular pigment, but also highly disease specific features can be imaged with FLIO. Furthermore, FLIO may be helpful to identify individuals with MacTel at very early stages of disease, possibly before they show clinical signs of retinal damage. This could help to initiate, once available, an early therapy. Overall, FLIO leads to improved clinical detection and diagnose of MacTel.

References

1. Charbel Issa P, et al. Macular telangiectasia type 2. Prog Retin Eye Res. 2013;34:49–77.
2. Clemons TE, et al. Baseline characteristics of participants in the natural history study of macular telangiectasia (MacTel) MacTel Project Report No. 2. Ophthalmic Epidemiol. 2010;17(1):66–73.
3. Finger RP, et al. Reading performance is reduced by parafoveal scotomas in patients with macular telangiectasia type 2. Invest Ophthalmol Vis Sci. 2009;50(3):1366–70.

4. Klein R, et al. The prevalence of macular telangiectasia type 2 in the Beaver Dam eye study. Am J Ophthalmol. 2010;150(1):55–62 e2.
5. Charbel Issa P, Holz FG, Scholl HP. Metamorphopsia in patients with macular telangiectasia type 2. Doc Ophthalmol. 2009;119(2):133–40.
6. Gass JDM. Stereoscopic atlas of macular diseases : diagnosis and treatment. 2nd ed. St. Louis: C. V. Mosby. xi; 1977. p. 411.
7. Gass JD, Blodi BA. Idiopathic juxtafoveolar retinal telangiectasis. Update of classification and follow-up study. Ophthalmology. 1993;100(10):1536–46.
8. Chin EK, et al. Staging of macular telangiectasia: power-Doppler optical coherence tomography and macular pigment optical density. Invest Ophthalmol Vis Sci. 2013;54(7):4459–70.
9. Helb HM, et al. Abnormal macular pigment distribution in type 2 idiopathic macular telangiectasia. Retina. 2008;28(6):808–16.
10. Delori FC, et al. Macular pigment density measured by autofluorescence spectrometry: comparison with reflectometry and heterochromatic flicker photometry. J Opt Soc Am A Opt Image Sci Vis. 2001;18(6):1212–30.
11. Zeimer MB, et al. Idiopathic macular telangiectasia type 2: distribution of macular pigment and functional investigations. Retina. 2010;30(4):586–95.
12. Degli Esposti S, et al. Macular pigment parameters in patients with macular telangiectasia (MacTel) and normal subjects: implications of a novel analysis. Invest Ophthalmol Vis Sci. 2012;53(10):6568–75.
13. Bernstein PS, et al. Lutein, zeaxanthin, and meso-zeaxanthin: the basic and clinical science underlying carotenoid-based nutritional interventions against ocular disease. Prog Retin Eye Res. 2016;50:34–66.
14. Choi RY, et al. Macular pigment distribution responses to high-dose Zeaxanthin supplementation in patients with macular telangiectasia Type 2. Retina. 2017;37:2238.
15. Li B, et al. Retinal accumulation of zeaxanthin, lutein, and beta-carotene in mice deficient in carotenoid cleavage enzymes. Exp Eye Res. 2017;159:123–31.
16. Powner MB, et al. Perifoveal muller cell depletion in a case of macular telangiectasia type 2. Ophthalmology. 2010;117(12):2407–16.
17. Powner MB, et al. Loss of Muller's cells and photoreceptors in macular telangiectasia type 2. Ophthalmology. 2013;120(11):2344–52.
18. Peto T, et al. CORRELATION OF CLINICAL AND STRUCTURAL PROGRESSION WITH VISUAL ACUITY LOSS IN MACULAR TELANGIECTASIA TYPE 2: MacTel Project Report No. 6-The MacTel Research Group. Retina. 2018;38(Suppl 1):S8–S13. https://doi.org/10.1097/IAE.0000000000001697. PMID:28505012.
19. Sallo FB, et al. The IS/OS junction layer in the natural history of type 2 idiopathic macular telangiectasia. Invest Ophthalmol Vis Sci. 2012;53(12):7889–95.
20. Sallo FB, et al. Retinal crystals in type 2 idiopathic macular telangiectasia. Ophthalmology. 2011;118(12):2461–7.
21. Ryu CL, et al. Macular Telangiectasia Type 2 in an otherwise healthy teenage boy with consanguineous parents. Retin Cases Brief Rep. 2018;12(3):200–3. https://doi.org/10.1097/ICB.0000000000000481. PMID: 27828905.
22. Baumuller S, et al. Outer retinal hyperreflective spots on spectral-domain optical coherence tomography in macular telangiectasia type 2. Ophthalmology. 2010;117(11):2162–8.
23. Davidorf FH, Pressman MD, Chambers RB. Juxtafoveal telangiectasis-a name change? Retina. 2004;24(3):474–8.
24. Abujamra S, et al. Idiopathic juxtafoveolar retinal telangiectasis: clinical pattern in 19 cases. Ophthalmologica. 2000;214(6):406–11.
25. Barthelmes D, Sutter FK, Gillies MC. Differential optical densities of intraretinal spaces. Invest Ophthalmol Vis Sci. 2008;49(8):3529–34.
26. Charbel Issa P, et al. Microperimetric assessment of patients with type 2 idiopathic macular telangiectasia. Invest Ophthalmol Vis Sci. 2007;48(8):3788–95.
27. Mehta H, et al. Natural history and effect of therapeutic interventions on subretinal fluid causing foveal detachment in macular telangiectasia type 2. Br J Ophthalmol.

2017;101(7):955–9. https://doi.org/10.1136/bjophthalmol-2016-309237. Epub 2016 Oct 28. PMID: 27793821.

28. Kovach JL, Rosenfeld PJ. Bevacizumab (avastin) therapy for idiopathic macular telangiectasia type II. Retina. 2009;29(1):27–32.

29. Roller AB, et al. Intravitreal bevacizumab for treatment of proliferative and nonproliferative type 2 idiopathic macular telangiectasia. Retina. 2011;31(9):1848–55.

30. Chew EY, et al. Ciliary neurotrophic factor for macular telangiectasia type 2: results from a phase 1 safety trial. Am J Ophthalmol. 2015;159(4):659–666 e1.

31. Bucher F, et al. CNTF attenuates vasoproliferative changes through upregulation of SOCS3 in a mouse-model of oxygen-induced retinopathy. Invest Ophthalmol Vis Sci. 2016;57(10):4017–26.

32. Parmalee NL, et al. Identification of a potential susceptibility locus for macular telangiectasia type 2. PLoS One. 2012;7(8):e24268.

33. Gillies MC, et al. Familial asymptomatic macular telangiectasia type 2. Ophthalmology. 2009;116(12):2422–9.

34. Delaere L, Spielberg L, Leys AM. Vertical transmission of macular telangiectasia type 2. Retin Cases Brief Rep. 2012;6(3):253–7.

35. Parmalee NL, et al. Analysis of candidate genes for macular telangiectasia type 2. Mol Vis. 2010;16:2718–26.

36. Scerri TS, et al. Genome-wide analyses identify common variants associated with macular telangiectasia type 2. Nat Genet. 2017;49(4):559–67.

37. Wong WT, et al. Fundus autofluorescence in type 2 idiopathic macular telangiectasia: correlation with optical coherence tomography and microperimetry. Am J Ophthalmol. 2009;148(4):573–83.

38. Niskopoulou M, et al. Is indocyanine green angiography useful for the diagnosis of macular telangiectasia type 2? Br J Ophthalmol. 2013;97(7):946–8.

39. Surguch V, Gamulescu MA, Gabel VP. Optical coherence tomography findings in idiopathic juxtafoveal retinal telangiectasis. Graefes Arch Clin Exp Ophthalmol. 2007;245(6):783–8.

40. Sallo FB, et al. "En face" OCT imaging of the IS/OS junction line in type 2 idiopathic macular telangiectasia. Invest Ophthalmol Vis Sci. 2012;53(10):6145–52.

41. Balaskas K, et al. Associations between autofluorescence abnormalities and visual acuity in idiopathic macular telangiectasia type 2: MacTel project report number 5. Retina. 2014;34(8):1630–6.

42. Heeren TF, et al. Progression of vision loss in macular Telangiectasia Type 2. Invest Ophthalmol Vis Sci. 2015;56(6):3905–12.

43. Schmitz-Valckenberg S, et al. Structural and functional changes over time in MacTel patients. Retina. 2009;29(9):1314–20.

44. Wu L, Evans T, Arevalo JF. Idiopathic macular telangiectasia type 2 (idiopathic juxtafoveolar retinal telangiectasis type 2A, Mac Tel 2). Surv Ophthalmol. 2013;58(6):536–59.

45. Sallo FB, et al. Multimodal imaging in type 2 idiopathic macular telangiectasia. Retina. 2015;35(4):742–9.

46. Toto L, et al. Multimodal imaging of macular telangiectasia Type 2: focus on vascular changes using optical coherence tomography angiography. Invest Ophthalmol Vis Sci. 2016;57(9):OCT268–76.

47. Sauer L, et al. Fluorescence lifetime imaging ophthalmoscopy: a novel way to assess macular Telangiectasia Type 2. Ophthalmol Retina. 2018;2(6):587–98.

48. Solberg Y, et al. Fluorescence lifetime patterns in macular Telangiectasia Type 2. Retina. 2019; https://doi.org/10.1097/IAE.0000000000002411. [Epub ahead of print]. PMID: 30664123.

49. Sauer L, et al. Impact of macular pigment on fundus autofluorescence lifetimes. Invest Ophthalmol Vis Sci. 2015;56(8):4668–79.

50. Sauer L, et al. Monitoring macular pigment changes in macular holes using Fluorescence Lifetime Imaging Ophthalmoscopy (FLIO). Acta Ophthalmol. 2017;95(5):481–92. https://doi.org/10.1111/aos.13269. Epub 2016 Oct 24. PMID: 27775222.

51. Schweitzer D. Metabolic mapping. In: Holz F, Spaide R, editors. Medical retina. Berlin/Heidelberg: Springer; 2010. p. 107–23.
52. Schweitzer D, et al. Towards metabolic mapping of the human retina. Microsc Res Tech. 2007;70(5):410–9.
53. Schweitzer D, et al. Fluorescence lifetime imaging ophthalmoscopy in type 2 diabetic patients who have no signs of diabetic retinopathy. J Biomed Opt. 2015;20(6):61106.
54. Schmidt J, et al. Fundus autofluorescence lifetimes are increased in non-proliferative diabetic retinopathy. Acta Ophthalmol. 2017;95(1):33–40. https://doi.org/10.1111/aos.13174. Epub 2016 Aug 13. PMID: 27519815.
55. Jentsch S, et al. Retinal fluorescence lifetime imaging ophthalmoscopy measures depend on the severity of Alzheimer's disease. Acta Ophthalmol. 2015;93(4):e241–7. https://doi.org/10.1111/aos.12609. Epub 2014 Dec 7. PMID: 25482990.
56. Klemm M, et al. Repeatability of autofluorescence lifetime imaging at the human fundus in healthy volunteers. Curr Eye Res. 2013;38(7):793–801.
57. Dysli C, et al. Fluorescence lifetime imaging in Stargardt disease: potential marker for disease progression. Invest Ophthalmol Vis Sci. 2016;57(3):832–41.
58. Dysli C, Wolf S, Zinkernagel MS. Autofluorescence lifetimes in geographic atrophy in patients with age-related macular degeneration. Invest Ophthalmol Vis Sci. 2016;57(6):2479–87.

Chapter 15
Hereditary Retinal Diseases: Stargardt, Choroideremia, Retinitis Pigmentosa

Chantal Dysli, Yasmin Solberg, and Lydia Sauer

Stargardt Disease

Stargardt disease (STGD) is the most common inherited macular dystrophy that causes visual impairment in childhood and young adults [1]. STGD is predominately inherited in an autosomal recessive trait and has been linked to mutations in the *ABCA4* gene [2]. *ABCA4* is an ATP-binding cassette of transmembrane rim proteins expressed in the outer segment discs of rod and cone photoreceptors, and is involved in the transport of all-trans-retinal. A defect in an *ABCA4* encoded rim protein leads to dysfunction of the normal visual cycle, with accumulation of all-trans-retinal in the photoreceptors and the RPE [3]. Subsequently, the oxidation and the buildup of all-trans retinaldehyde induces the formation of N-retinylidene-N-retinyl-ethanolamine (A2E), a component of lipofuscin [4, 5]. The characteristic phenotypic features, such as the pisciform flecks and atrophy, are thought to develop as a result of the accumulation of lipofuscin and retinoid by-products in the RPE [6]. Histologically, flecks are located at the level of the RPE. Over the course of the disease, initially well-defined flecks have been shown to progress outwards from the central macular in a centrifugal pattern. They may reabsorb, leaving poorly demarcated lesions and residual atrophy resulting in a "beaten-bronze" appearance [7–9]. Thus, at advanced stages of disease, a progressive bilateral atrophy of the RPE,

C. Dysli (✉)
Department of Ophthalmology and Department of Clinical Research, Inselspital, Bern University Hospital, University of Bern, Bern, Switzerland
e-mail: chantal.dysli@insel.ch

Y. Solberg
Department of Ophthalmology, University Hospital Zurich, Zurich, Switzerland

L. Sauer
Department of Ophthalmology and Visual Sciences, Moran Eye Center, University of Utah, Salt Lake City, UT, USA

© Springer Nature Switzerland AG 2019
M. Zinkernagel, C. Dysli (eds.), *Fluorescence Lifetime Imaging Ophthalmoscopy*, https://doi.org/10.1007/978-3-030-22878-1_15

photoreceptors, and choroidal vasculature can be found [10]. However, due to the large number of disease-causing sequence variants identified in *ABCA4* (>1000), the characteristics of STGD varies widely from patient to patient, and severity of fundus abnormalities seen in ophthalmoscopy often are not directly associated with the visual acuity [11–13]. Fishman et al. characterized STGD into four phenotypic subtypes based on clinical appearance and electrophysiological findings [14]. In addition to clinical evaluation, several imaging techniques have been established in the diagnosis, characterization, monitoring, and investigation of the underlying pathophysiology of STGD, including FA, FAF, spectral domain OCT, and electro-retinography (ERG).

FLIO measurements in STGD display characteristic fluorescence lifetime patterns, which differ from the FLIO pattern found in healthy eyes [15–18]. Lifetimes within areas of intact retina without retinal deposits or atrophic lesions are comparable to age-matched healthy controls. However, within hyperfluorescent flecks seen on autofluorescence intensity images in the area of retinal deposits in OCT, both, shorter and longer fluorescence lifetime values compared to the surrounding retina were observed. Primarily, flecks displayed long (blue) lifetimes, however, in 91% of the investigated eyes a small number of flecks with shorter lifetimes, color coded in red, are visible (Fig. 15.1). The short lifetime flecks were more apparent in the LSC. Correlation to their corresponding intensity images revealed that not all short lifetime flecks were visible on intensity images. In follow-up examinations these short lifetime lesions became visible in the FAF intensity images, and progressed to flecks with characteristic long lifetimes in FLIO over time. Figure 15.2 shows follow-up measurements after 6 and 32 months. In the transition phase from short to long lifetimes, longer lifetimes initiate in the center of the flecks, and radiate outwards with time. Interestingly, patients seem to report visual changes only when presenting with long FLIO lifetime flecks but not with the initial flecks that show shortened FLIO lifetimes. These changing autofluorescence lifetime patterns may reflect intracellular events in RPE cells and remodelling of the flecks. There is strong evidence that STGD is primarily a disease of the outer neural retina, with the RPE being affected secondarily [8]. Short autofluorescence lifetimes may originate from degenerating photoreceptor cells containing bis-retinoid fluorophores and retinaldehyde adducts [19]. Long lifetimes could be explained by accumulation of autofluorescent precursors, which may contribute to RPE- and outer photoreceptor segment dysfunction, leading to lipofuscin build up with the accumulation of photo-oxidation and photo-degradation products of A2E compounds [20]. In the late stage of disease, atrophic areas seen in FAF intensity images generally exhibit significantly prolonged mean fluorescence lifetimes compared to the surrounding unaffected retina in both spectral channels. However, areas of short fluorescence lifetimes may still be present. A prolongation of fluorescence lifetimes with remaining areas of short lifetimes, depending on the degree of photoreceptor degradation, was also described in geographic atrophy in late age-related macular degeneration [21].

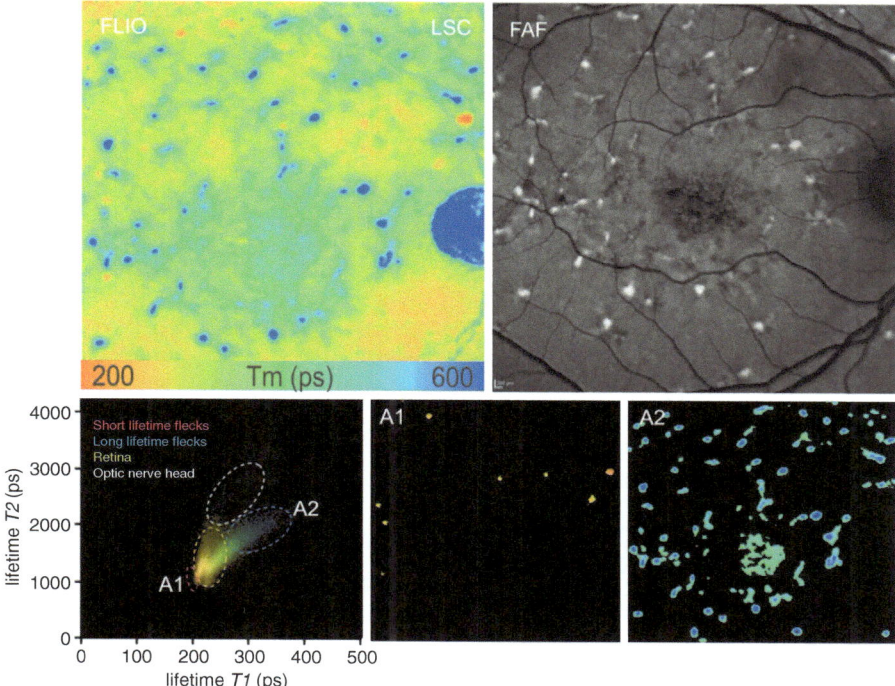

Fig. 15.1 FLIO in Stargardt disease. Fluorescence lifetime image (FLIO, long spectral channel) and, fundus autofluorescence intensity (FAF) image, of a patient with Stargardt disease. Hyperfluorescent flecks may feature shorter or longer fluorescence lifetimes compared to the surrounding retina. Corresponding 2D analysis with distribution histograms of the short and the long fluorescence lifetime components *T1* and *T2* (bottom). Specific areas are highlighted according to the lifetime distribution clouds: surrounding retina, short fluorescence lifetime flecks, long lifetime flecks, and optic nerve head

To date, there are no treatments available for STGD. The only preventive measures for slowing down disease progression include sunlight protection and a normal diet without inappropriate prescription of vitamin A supplements. However, three paths of intervention are currently being explored in human clinical trials, including stem cell therapy, gene replacement therapy, and pharmacological approaches, such as deuterated vitamin A [22, 23].

FLIO likely highlights early disease-related changes, and may therefore be used as a screening modality for disease activity. It might also be an appropriate tool for detecting subtle retinal changes over time, and could be useful in the development of new outcome measurements for clinical trials testing novel therapies for STGD. Possibly, such trials should be targeted to patients with flecks that show shortened FLIO lifetimes, as these flecks are likely still developing, and vision loss in these areas may be preventable.

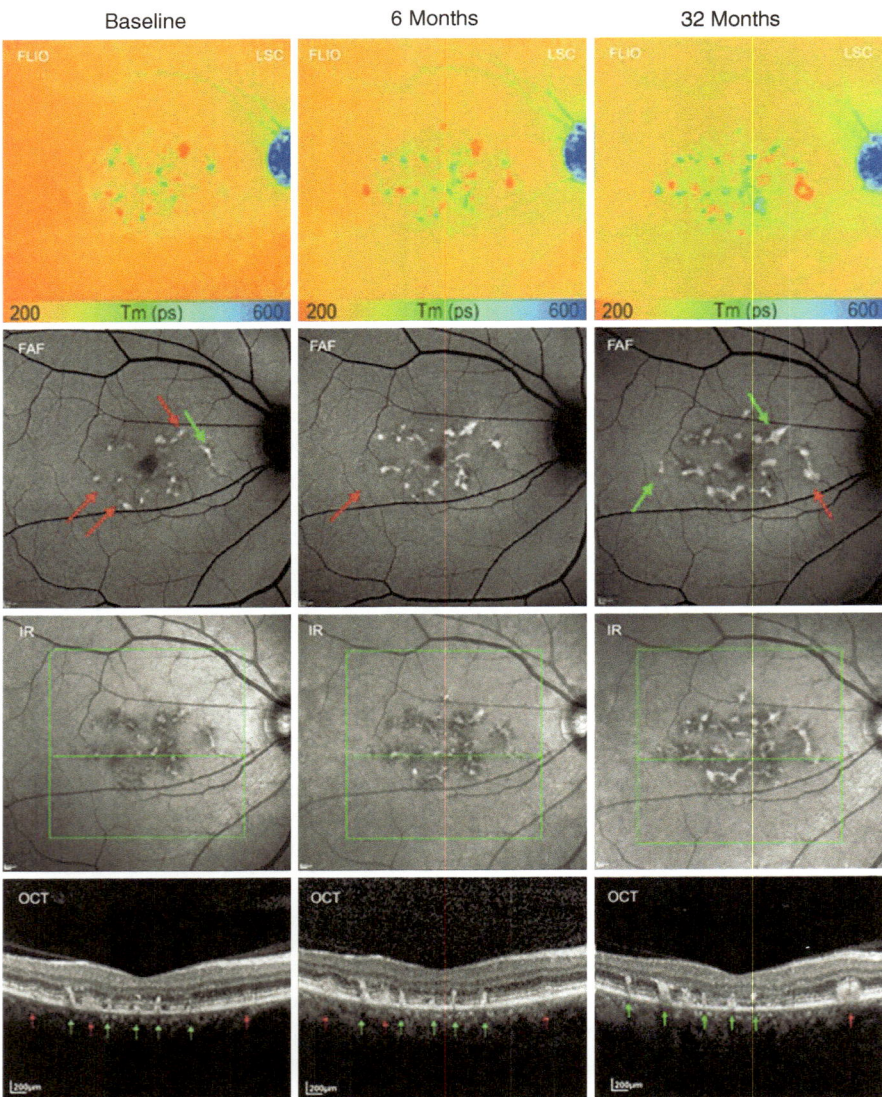

Fig. 15.2 Disease progression of Stargardt disease within 6 and 32 months follow-up. Fluorescence lifetime (FLIO, long spectral channel) and fundus autofluorescence intensity (FAF) images are shown. Correlating optical coherence tomography scans (OCT) of the indicated lines in the infrared (IR) images are shown below. From baseline (left) to 6 months follow-up (middle), to 32 months follow-up (right), clear disease progression is visible with transition of flecks with short fluorescence lifetimes (red arrows) to flecks with long fluorescence lifetimes (green arrows) and appearance of new hyperfluorescent flecks

Choroideremia

Choroideremia (CHM) is a rare hereditary degenerative disease of the choroid, the RPE, and the neurosensory retina with an estimated prevalence of 1:50,000 people of European descent [24]. It is inherited in an X-linked, recessive pattern and thereby mainly affects young male subjects. Disease causing is a mutation in the CHM gene on the long arm of the X chromosome, coding for Rab escort protein-1 (REP 1) which is responsible for the membrane trafficking in the retina and the RPE. Clinically, extensive chorioretinal atrophy with characteristic pallor of the fundus due to increased transilluminescence of the sclera can be seen on the posterior pole with progression from the periphery towards the macular center. Patients may report night blindness as one of the first symptoms due to peripheral loss of rods and underlaying retinal layers. In early stages of disease, the central visual acuity is mostly unaffected and may still be preserved till up to the age of 50 to 70 years. However, remaining retinal islands shrink over time, and choroideremia may end in total blindness. Imaging methods such as OCT, infrared- and autofluorescence imaging rendered as very useful methods for diagnosis and follow-up examinations in choroideremia. Additionally, genetic testing is used to confirm the diagnosis. By now, there is no curable standard therapy available. However, several clinical trials are ongoing, including gene replacement therapies using adeno-associated virus as vectors for replacement of the defective gen sequence [25]. Currently, mainly patients with advanced disease stage were included in such clinical trials. However, future aims are to include earlier stages of disease in order to prevent further damage, and preserve visual acuity and visual field over time. Therefore, corresponding imaging modalities for identification of suitable subjects as well as for monitoring of subtle retinal changes are much needed.

FAF intensity measurement well describes areas of intact RPE which appears as bright due to normal autofluorescence of the lipofuscin in the RPE. In contrast, surrounding retina with RPE atrophy appears as hypoautofluorescent. FLIO imaging in a study including 16 eyes of eight patients with advanced CHM revealed much prolonged fluorescence lifetimes in areas of complete RPE- and photoreceptor atrophy (Fig. 15.3) [26]. Shortest lifetimes are still measured in the macular center with preserved retinal layer structure. However, in areas of RPE atrophy but remaining photoreceptor layers, still short fluorescence lifetimes can be measured. They may origin from remaining visual cycle activity and accumulation of visual cycle by-products. In follow-up examinations, a decrease of short fluorescence lifetimes correlated with disease progression and increase of chorioretinal atrophy.

In CHM, FLIO represents a reproducible method to image and follow-up subtle retinal changes of the RPE and the photoreceptors. Thereby, the technique might provide a useful imaging tool regarding monitoring of therapeutical interventions for differentiated readouts.

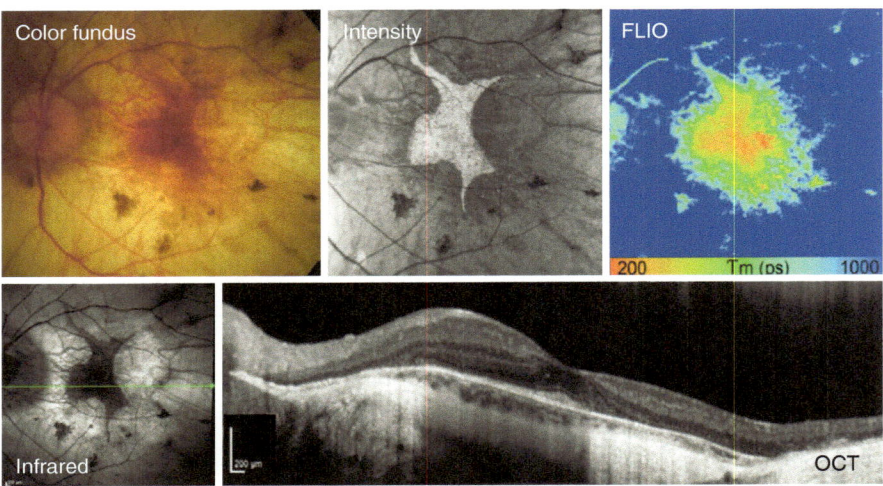

Fig. 15.3 Multimodal imaging in a patient with choroideremia. The borders of the RPE atrophy is best seen in the autofluorescence intensity image. Areas of atrophy of the RPE and the outer retinal layers featured the longest lifetimes, whereas areas of preserved photoreceptor layers in the absence of the RPE featured intermediate lifetimes

Retinitis Pigmentosa

Retinitis pigmentosa (RP) is an inherited, degenerative eye disease leading to peripheral retinal degeneration. It can show little progression over decades, but may also escalate to legal blindness [27, 28]. RP is a heterogenous diagnosis, as a vast number of etiologies from at least 79 different genes contribute to the development of this disease [29, 30]. These different genotypes often result in similar phenotypes, where the peripheral retina degenerates but the central macula is relatively spared in many cases. The inheritance model of RP is relatively diverse, approximately 30–40% of genes are autosomal dominant, 50–60% of genes are autosomal recessive, and about 5–15% of the cases show an X-linked inheritance pattern [27]. The onset of disease is usually in childhood or early adulthood, and often preceded by nyctalopia [31, 32]. A progressive peripheral vision loss with intact central vision is characteristic for RP, visual fields show a tunnel vision that constricts over time [33]. Interestingly, however, the function of retinal photoreceptors was reported to decrease even before patients become symptomatic, and before RPE atrophy manifests [34, 35]. Macular pigment levels in patients with spared central vision appear comparable to those in healthy controls [36].

Recently, two studies performed in two different centers independently focused on describing FLIO lifetimes in RP, and both studies showed a specific disease-related pattern [36, 37]. Ring-shaped features in FLIO appear to correspond to different areas of the disease manifestation. Figure 15.4a–d shows typical FLIO patterns in eyes with RP. Areas of peripheral RPE- and photoreceptor atrophy were identified to show prolonged FLIO lifetimes (LSC: around 400 ps) as compared to age-matched healthy controls (LSC: around 280 ps). This was a highly significant

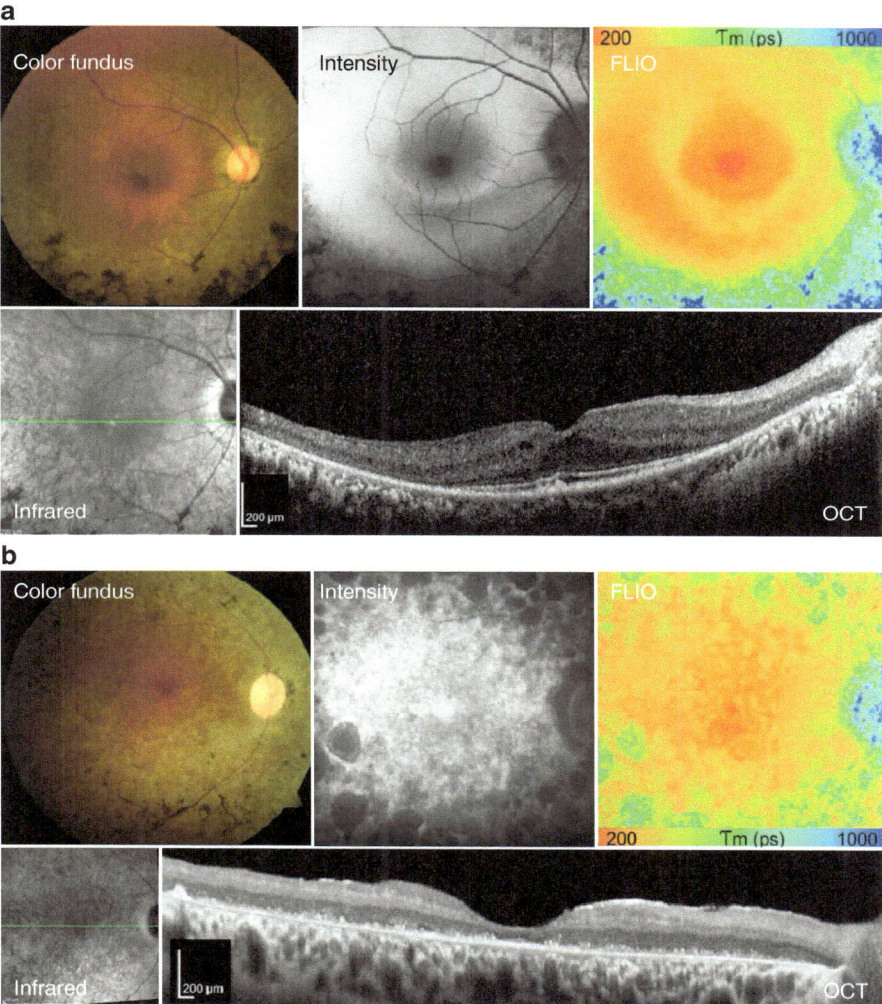

Fig. 15.4 (**a–d**) Multimodal imaging in patients with retinitis pigmentosa. Different disease stages feature individual FLIO pattern. The ring shaped pattern is clearly seen in the FLIO images compared to other imaging modalities

difference (p < 0.001) [36]. However, areas with remaining photoreceptors within underlying RPE atrophy were found to show only slightly prolonged FLIO lifetimes (around 300 ps) compared to healthy eyes.

The ring-like structures were further investigated in both FLIO studies, and good correlation with functional imaging and visual acuity were reported [36, 37]. FLIO thereby confirms prior studies that used FAF intensity imaging and described a parafoveal hyperfluorescent ring with coincident peripheral hypofluorescence, the latter being indicative of retinal atrophy [38]. In the course of disease progression, the visual fields shrink in a similar way as the rings [39].

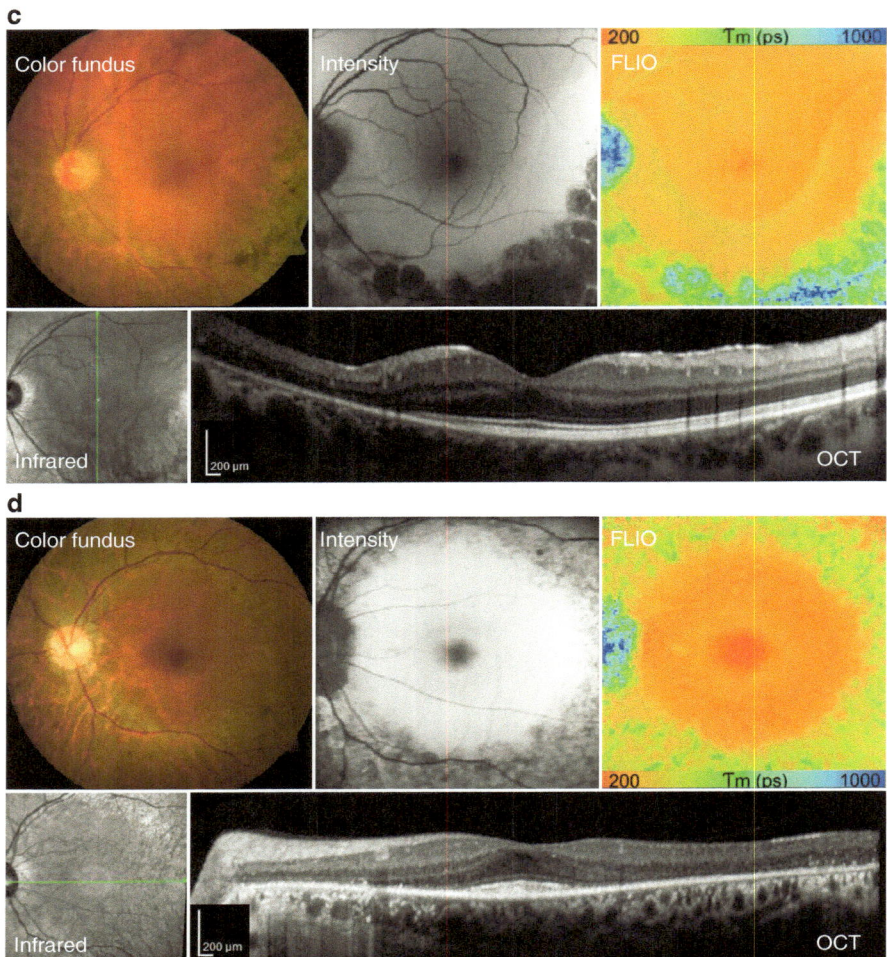

Fig. 15.4 (continued)

FLIO may be useful for characterization of phenotypical patterns in the various genetic subtypes of RP [36]. Hyperfluorescent rings appear as an area of generally short lifetimes, and long lifetimes can be found in peripheral atrophic regions. An inheritance-dependent phenotypic variance was found, which was related to the prominence of the ring-like patterns. Patients with autosomal dominant RP as well as Usher syndrome showed the strongest ring-like patterns. Patients with autosomal recessive RP presented with a milder pattern, and patients with X-linked RP did not show a ring-like pattern at all [40].

Both FLIO studies were able to highlight that central areas of short FLIO lifetimes correspond to MP, which remains relatively unaffected in RP. However, shorter FLIO lifetimes in the foveal center correlated with better visual acuity. FLIO may give additional insights in the disease process as well as the integrity of the retina in RP. It may be useful in the analysis of different genetic subtypes, their pathology and pathogenesis, as well as the basic principles behind disease progression [36, 37].

Summary Box

FLIO shows characteristic and disease specific fluorescence lifetime patterns in different hereditary retinal diseases. Thereby, the lifetime component of retinal autofluorescence measurement provides additional information compared to fundus autofluorescence intensity measurement, and might indicate different components of deposits and configuration of the retinal layer structure in congenital retinopathies. Due to a high sensitivity for subtle retinal changes, FLIO might provide a sensitive tool for diagnosis and follow-up for natural disease progression and for clinical trials.

References

1. Walia S, Fishman GA. Natural history of phenotypic changes in Stargardt macular dystrophy. Ophthalmic Genet. 2009;30(2):63–8.
2. Haji Abdollahi S, Hirose T. Stargardt-Fundus flavimaculatus: recent advancements and treatment. Semin Ophthalmol. 2013;28(5–6):372–6.
3. Koenekoop RK. The gene for Stargardt disease, ABCA4, is a major retinal gene: a mini-review. Ophthalmic Genet. 2003;24(2):75–80.
4. Allikmets R, et al. Mutation of the Stargardt disease gene (ABCR) in age-related macular degeneration. Science. 1997;277(5333):1805–7.
5. Allikmets R. A photoreceptor cell-specific ATP-binding transporter gene (ABCR) is mutated in recessive Stargardt macular dystrophy. Nat Genet. 1997;17(1):122.
6. Smith RT, et al. Lipofuscin and autofluorescence metrics in progressive STGD. Invest Ophthalmol Vis Sci. 2009;50(8):3907–14.
7. Cukras CA, et al. Centrifugal expansion of fundus autofluorescence patterns in Stargardt disease over time. Arch Ophthalmol. 2012;130(2):171–9.
8. Gomes NL, et al. A comparison of fundus autofluorescence and retinal structure in patients with Stargardt disease. Invest Ophthalmol Vis Sci. 2009;50(8):3953–9.
9. Armstrong JD, Xu S, Elfervig JL. Long-term follow-up of Stargardt's disease and fundus flavimaculatus. Ophthalmology. 1998;105:448–57.
10. Rotenstreich Y, Fishman GA, Anderson RJ. Visual acuity loss and clinical observations in a large series of patients with Stargardt disease. Ophthalmology. 2003;110(6):1151–8.
11. Fishman GA, et al. Variation of clinical expression in patients with Stargardt dystrophy and sequence variations in the ABCR gene. Arch Ophthalmol. 1999;117(4):504–10.
12. Fishman GA, et al. Visual acuity loss in patients with Stargardt's macular dystrophy. Ophthalmology. 1987;94(7):809–14.
13. Cornelis SS, et al. In Silico functional meta-analysis of 5,962 ABCA4 variants in 3,928 retinal dystrophy cases. Hum Mutat. 2017;38(4):400–8.
14. Fishman GA. Fundus flavimaculatus. A clinical classification. Arch Ophthalmol. 1976;94(12):2061–7.
15. Dysli C, et al. Fluorescence lifetime imaging in Stargardt disease: potential marker for disease progression. Invest Ophthalmol Vis Sci. 2016;57(3):832–41.
16. Dysli C, et al. Quantitative analysis of fluorescence lifetime measurements of the macula using the fluorescence lifetime imaging ophthalmoscope in healthy subjects. Invest Ophthalmol Vis Sci. 2014;55(4):2106–13.

17. Dysli C, et al. Fluorescence lifetime imaging ophthalmoscopy. Prog Retin Eye Res. 2017;60:120–43.
18. Solberg Y, Dysli C, Escher P, Berger L, Wolf S, Zinkernagel MS. Retinal flecks in Stargardt Disease reveal characteristic fluorescence lifetime transition over time. Retina. 2019;39(5):1. https://doi.org/10.1097/IAE.0000000000002519.
19. Sparrow JR, et al. Flecks in recessive Stargardt disease: short-wavelength autofluorescence, near-infrared autofluorescence, and optical coherence tomography. Invest Ophthalmol Vis Sci. 2015;56(8):5029–39.
20. Liu J, et al. The biosynthesis of A2E, a fluorophore of aging retina, involves the formation of the precursor, A2-PE, in the photoreceptor outer segment membrane. J Biol Chem. 2000;275(38):29354–60.
21. Dysli C, Wolf S, Zinkernagel MS. Autofluorescence lifetimes in geographic atrophy in patients with age-related macular degeneration. Invest Ophthalmol Vis Sci. 2016;57:2479–87. https://doi.org/10.1167/iovs.15-18381.
22. Charbel Issa P, Barnard AR, Herrmann P, Washington I, MacLaren RE. Rescue of the Stargardt phenotype in Abca4 knockout mice through inhibition of vitamin A dimerization. Proc Natl Acad Sci U S A. 2015;112(27):8415–20.
23. Campa C, et al. The role of gene therapy in the treatment of retinal diseases: a review. Curr Gene Ther. 2017;17(3):194–213.
24. Zinkernagel MS, MacLaren RE. Recent advances and future prospects in choroideremia. Clin Ophthalmol. 2015;9:2195–200.
25. Xue K, et al. Beneficial effects on vision in patients undergoing retinal gene therapy for choroideremia. Nat Med. 2018;24(10):1507–12.
26. Dysli C, et al. Autofluorescence lifetimes in patients with choroideremia identify photoreceptors in areas with retinal pigment epithelium atrophy. Invest Ophthalmol Vis Sci. 2016;57(15):6714–21.
27. Hartong DT, Berson EL, Dryja TP. Retinitis pigmentosa. Lancet. 2006;368(9549):1795–809.
28. Shankar S. Hereditary retinal and choroidal dystrophies. In: Rimoin D, Pyeritz R, Korf B, editors. Emery and Rimoin's principles and practice of medical genetics. London: Academic Press; 2013. p. 1–18. see also: https://www.elsevier.com/books/emery-and-rimoins-principles-and-practice-of-medical-genetics/rimoin/978-0-12-383834-6.
29. Ferrari S, et al. Retinitis pigmentosa: genes and disease mechanisms. Curr Genomics. 2011;12(4):238–49.
30. Zhang Q. Retinitis pigmentosa: progress and perspective. Asia Pac J Ophthalmol (Phila). 2016;5(4):265–71.
31. Daiger SP, Bowne SJ, Sullivan LS. Perspective on genes and mutations causing retinitis pigmentosa. Arch Ophthalmol. 2007;125(2):151–8.
32. Fishman GA. Retinitis pigmentosa. Visual loss. Arch Ophthalmol. 1978;96(7):1185–8.
33. Hamel C. Retinitis pigmentosa. Orphanet J Rare Dis. 2006;1:40.
34. Berson EL. Retinitis pigmentosa. The Friedenwald lecture. Invest Ophthalmol Vis Sci. 1993;34(5):1659–76.
35. Murakami T, et al. Association between abnormal autofluorescence and photoreceptor disorganization in retinitis pigmentosa. Am J Ophthalmol. 2008;145(4):687–94.
36. Andersen K, Sauer L, et al. Characterization of retinitis pigmentosa using Fluorescence Lifetime Imaging Ophthalmoscopy (FLIO). Transl Vis Sci Technol. 2018;7:20.
37. Dysli C, et al. Fundus autofluorescence lifetime patterns in retinitis pigmentosa. Invest Ophthalmol Vis Sci. 2018;59(5):1769–78.
38. Lima LH, et al. Structural assessment of hyperautofluorescent ring in patients with retinitis pigmentosa. Retina. 2009;29(7):1025–31.
39. Aizawa S, et al. Changes of fundus autofluorescence, photoreceptor inner and outer segment junction line, and visual function in patients with retinitis pigmentosa. Clin Exp Ophthalmol. 2010;38(6):597–604.
40. Andersen KM, et al. Characterization of retinitis pigmentosa using Fluorescence Lifetime Imaging Ophthalmoscopy (FLIO). Transl Vis Sci Technol. 2018;7(3):20.

Chapter 16
Macular Pigment

Lydia Sauer and Paul S. Bernstein

Retinal Carotenoids

Macular carotenoids, or macular pigment (MP), are dietary supplements that are assumed to protect the eye from light damage as well as retinal diseases, such as age-related macular degeneration [1]. Carotenoids are hydrophobic compounds that cannot be synthesized in the human body. MP accumulates at the fovea with a 1:1:1 mixture of the three different xanthophylls lutein (L), zeaxanthin (Z), and *meso*-zeaxanthin (MZ). L and Z are taken up by diet, whereas L can then be converted to MZ by RPE65 convertase [2]. MP is highly concentrated in the fovea, especially within the Müller cells of the Henle fiber layer. Two carotenoid-binding proteins are responsible for the specific distribution of MP in the eye. The zeaxanthin-binding protein (GSTP1) and lutein-binding protein (StARD3) have been identified as the reason for this spatial distribution inside the human retina [3, 4].

Like other carotenoids, the macular carotenoids are natural antioxidants. MP quenches free radicals and also absorbs blue light before it reaches the photoreceptor layer. The short wavelength of blue light exhibits high energy and is especially damaging to the retina [5–7]. Lower carotenoid status is associated with retinal diseases, and carotenoid supplementation is thought to help reduce the risk or delay the progression of intermediate AMD [8]. Therefore, an accurate non-invasive assessment of carotenoid status can help ophthalmologists identify patients most likely to benefit from carotenoid supplementation. Intensive research has resulted in a variety of innovative techniques for carotenoid assessment, which were recently evaluated [9]. It is difficult to point out a preferred technique, as all have advantages and disadvantages. FLIO emerges as a promising tool, as in contrast to other objective measurement modalities, FLIO does not rely on healthy surrounding areas in order to calculate the amount of MP. FLIO as a method to describe and quantify MP was first described in

L. Sauer (✉) · P. S. Bernstein
University of Utah, John A. Moran Eye Center, Salt Lake City, UT, USA
e-mail: Lydia.Sauer@hsc.utah.edu

© Springer Nature Switzerland AG 2019
M. Zinkernagel, C. Dysli (eds.), *Fluorescence Lifetime Imaging Ophthalmoscopy*, https://doi.org/10.1007/978-3-030-22878-1_16

2015 [10–13]. Before that, it was believed that MP only absorbs light but does not show fluorescence in the retina. These assumptions were made because of the very weak fluorescence intensity of MP, which manifests as a hypofluorescent (dark) spot in FAF intensity images. However, the fluorescence of carotenoids *in vivo* was first described using resonance Raman Spectroscopy [14]. Using FLIO, the fluorescence of carotenoids was confirmed in clinical studies [11, 15]. The principle that fluorescence lifetimes are independent of the fluorescence intensity is the main factor that allows for a detection of MP with this method [16]. Although weak in intensity, retinal carotenoids show short autofluorescence lifetimes. In a first study, 48 young and healthy subjects underwent FLIO as well as MP measurements [11]. The MP readings in FLIO correlated strongly and inversely with individual amounts of MP [11]. This was later confirmed by a second study that used a different method to assess MP amounts [15]. Here, 31 healthy subjects were included, and the correlation of MP with short foveal fluorescence lifetimes was confirmed. Figure 16.1 shows different MP distribution variances in healthy eyes, imaged with both FLIO as well as dual wavelength autofluorescence imaging. Beyond that, *ex vivo* studies determined the individual lifetimes of L and Z to be around 50–60 ps [15]. The benefits of utilizing FLIO in studies about MP includes the minimal effort required by the patient, the non-invasiveness, the short measurement times (around 2 minutes), as well as an image that does not require healthy reference regions to calculate MP. Therefore, FLIO can be used to detect MP in patients with abnormal lipofuscin distributions as well as patients with albinism [15].

FLIO correlates strongly with different modalities to evaluate the amount and distribution of MP, such as single-wavelength reflectometry and dual-wavelength AFI. It therefore proves suitable to be used as an additional tool for investigation of MP and detection of corresponding retinal diseases [17–19].

Carotenoid Measurement Ex Vivo

Crystals of Z and L were dissolved in CHAPS-PBS and measured *ex vivo* in 1–2 mm path-length quartz cuvettes, which were then placed in a special FLIO cuvette holder attachment. The carotenoids were measured in free state as well as bound to their respective binding proteins. Two different acquisition times (2 and 10 minutes) were used; the average photon counts were recorded as well. Mean autofluorescence lifetimes from each measurement were obtained from a rectangular region of the cuvette.

At various concentrations and with different durations, carotenoids consistently showed very short mean FAF lifetimes in the SSC (L: around 50 ps; and Z: around 60 ps). The cuvette as well as the solvent (CHAPS-PBS) did not show any measurable fluorescence. Complexed with their respective specific binding proteins in 1–5% methanol, the mean autofluorescence lifetimes prolonged. The long spectral channel's carotenoid fluorescence was extremely weak and therefore not investigated in this study. This is in accordance with reported fluorescence emission spectra of carotenoids, which predominantly are found within the SSC [14].

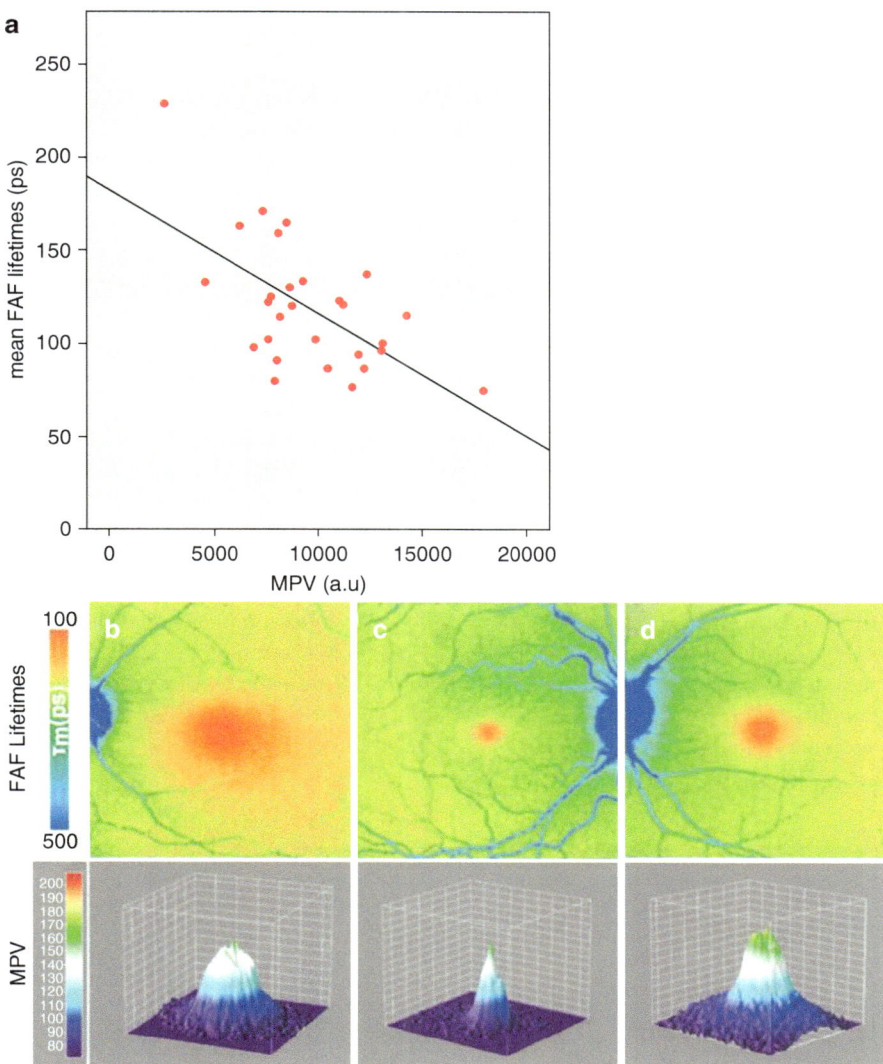

Fig. 16.1 FLIO imaging allows for detection and quantification of macular pigment distribution within the posterior pole. (**a**) Mean fluorescence lifetimes form the foveal center correlate well with the macular pigment volume (MPV, r = −0.6). FLIO images and macular pigment plots from three individuals with different distributions of macular pigment are shown (**b-d**)

The *ex vivo* results confirm the short mean autofluorescence lifetimes of carotenoids, which can be detected with the FLIO technology. The detected lifetimes for carotenoids likely present the resolution limit of the FLIO camera [11]. This knowledge about carotenoid fluorescence characteristics allows for a better understanding of the physical state of MP in the human fovea.

Macular Holes

Macular holes (MH) are defects in the center of the fovea, the most common form is age-related and idiopathic. Patients typically complain about visual disturbances on the affected eye. MH are reported with a prevalence of 3 in 1000 people, mainly affect patients in their sixth or seventh life decade, and occur more often in women than in men. Vitreo-foveal tractions seem to be causative [20]. Clinical diagnosis primarily uses OCT imaging and effective treatment is established with vitrectomy and gas/oil tamponade [21, 22]. 80% to 90% of eyes treated with surgery show an improvement in visual acuity [23].

MH present as an excellent model to investigate the spatial distribution of MP. In this disease, different retinal layers can be disrupted, and in combination with OCT imaging, different layers at the human fundus can be compared to corresponding FLIO signals. FLIO imaging shows a distribution of short FAF lifetimes in a ring-like distribution around the area of the macular hole. Figure 16.2 shows a typical FLIO image of an eye with a MH. This proves that MP is dislocated around the defect in a similar way as the inner retinal layers are dislocated [24]. Disruption of the Henle fiber layer seems to be critical for the presence or absence of short FAF lifetimes [12]. However, the presence of short FAF lifetimes despite the disruption of the Henle fiber layer confirms that MP is not solely restricted to this layer. Additionally, many MH contain a so-called operculum, a small opacity above the foveal hole [25]. It was shown that these opercula contain MP [26, 27]. In FLIO imaging, short lifetimes could also be found in such opercula [12].

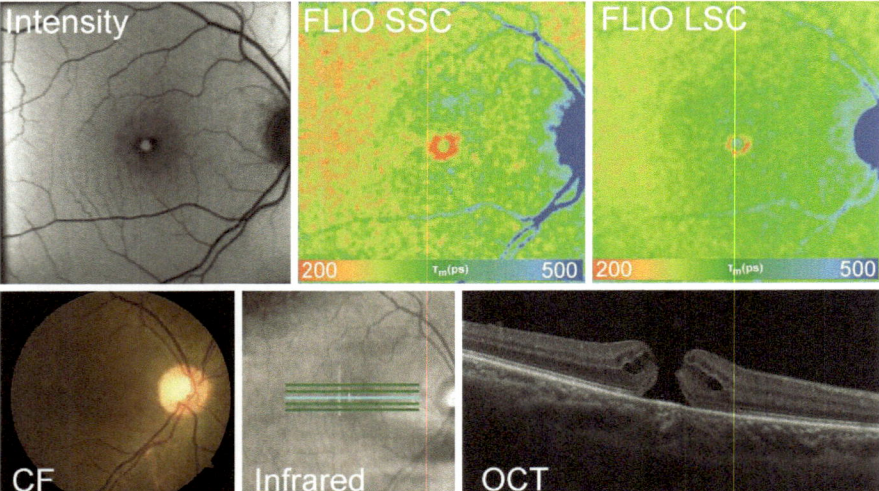

Fig. 16.2 FLIO imaging in a patient with a full thickness macular whole. FAF intensity and FLIO lifetime maps show abnormal distribution of the macular pigment, which in this patient is located adjacent to the macular hole

Imaging MP in MH help to understand the basic distribution patterns of carotenoids, however, it may also indicate visual improvement with surgery. Studies found that MP re-distributes towards the center of the fovea with successful surgery which seemed to correlate with better visual outcomes [12].

Albinism

Albinism presents with the absence of any pigments in the body, as melanocytes are unable to produce melanin. Melanin normally protects the cells from ultra-violet radiation from sunlight, therefore patients with albinism are prone to light damages. It is commonly known that albinism causes a less pigmented iris, which leads to red-appearing eyes as the red from the retina is apparent through the iris. However, changes with albinism go beyond that, and may affect the retina itself. Patients with albinism often present with reduced foveal depression [1, 28–31]. This finding can occur without typical skin findings, a disease called ocular albinism. The MP levels in patients with albinism are typically very low or not even detectable, but the extent of measurable MP in albinism has been discussed diversely [30, 32, 33].

Multiple patients with albinism were investigated with FLIO, a recent study showed two different individuals [34]. The patients presented in this study were only mildly affected and still had a good visual acuity and no nystagmus. Further measurements on patients with more severe forms show that FLIO measurements are also possible in patients with nystagmus due to albinism, Fig. 16.3 presents one example of such a patient.

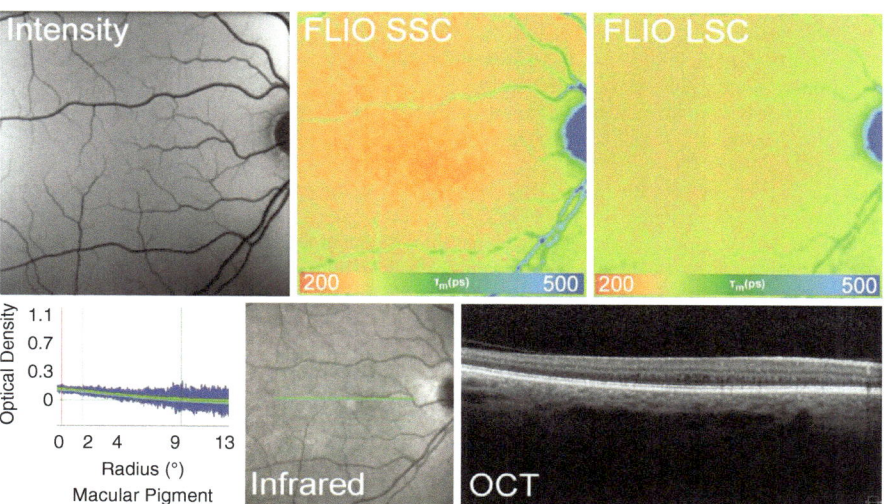

Fig. 16.3 FLIO imaging in a subject with albinism. The lack of macular pigment leads to an absence of hypoautofluorescence in the intensity image. Corresponding areas in the FLIO images show absence of the central short FLIO lifetimes, generally depicted in red color. The macular pigment measurement confirms the absence of a central macular pigment peak. The OCT illustrates the foveal hyperplasia

In FAF intensity images, the posterior pole of these patients seems to have normal autofluorescence with the exception of little to no MP absorbance. FLIO imaging shows no short FAF lifetimes from the center of the fovea. Only one eye of one patient with mild ocular albinism showed little amounts of MP. A previous study investigating MP in patients with albinism using with dual-wavelength AFI suggested evidence for MP accumulation [32]. However, other reports did not confirm this [30, 33]. It was described that dual-wavelength AFI often shows a sloping baseline of MP, which may be a miscalculation due to the absence of melanin. FLIO can confirm that there is no MP measurable in these patients [34]. This again confirms that MP impacts the short FAF lifetimes in the foveal center. Thereby, FLIO may be useful in imaging patients with potential albinism.

Summary Box

FLIO detects fluorescence signals from the macular carotenoids within the fovea. Macular pigment shows short FLIO lifetimes. In different retinal diseases with reduced or altered macular pigment, FLIO can provide additional information on the carotenoid status of patients.

References

1. Bernstein PS, et al. Lutein, zeaxanthin, and meso-zeaxanthin: the basic and clinical science underlying carotenoid-based nutritional interventions against ocular disease. Prog Retin Eye Res. 2016;50:34–66.
2. Shyam R, et al. RPE65 has an additional function as the lutein to meso-zeaxanthin isomerase in the vertebrate eye. Proc Natl Acad Sci U S A. 2017;114(41):10882–7.
3. Bhosale P, et al. Identification and characterization of a Pi isoform of glutathione S-transferase (GSTP1) as a zeaxanthin-binding protein in the macula of the human eye. J Biol Chem. 2004;279(47):49447–54.
4. Li B, et al. Identification of StARD3 as a lutein-binding protein in the macula of the primate retina. Biochemistry. 2011;50(13):2541–9.
5. Ham WT Jr, et al. Histologic analysis of photochemical lesions produced in rhesus retina by short-wave-length light. Invest Ophthalmol Vis Sci. 1978;17(10):1029–35.
6. Krinsky NI. Antioxidant functions of carotenoids. Free Radic Biol Med. 1989;7(6):617–35.
7. Bone RA, et al. Distribution of lutein and zeaxanthin stereoisomers in the human retina. Exp Eye Res. 1997;64(2):211–8.
8. Akuffo KO, et al. Relationship between macular pigment and visual function in subjects with early age-related macular degeneration. Br J Ophthalmol. 2017;101(2):190–7. https://doi.org/10.1136/bjophthalmol-2016-308418. Epub 2016 Apr 18. PMID: 27091854.
9. Sauer LLB, Bernstein PS. Ocular carotenoid status in health and disease. Annu Rev Nutr. 2019; https://doi.org/10.1146/annurev-nutr-082018-124555. [Epub ahead of print]. PMID: 31091415.
10. Schweitzer D, et al. Time-correlated measurement of autofluorescence. A method to detect metabolic changes in the fundus. Ophthalmologe. 2002;99(10):774–9.

11. Sauer L, et al. Impact of macular pigment on fundus autofluorescence lifetimes. Invest Ophthalmol Vis Sci. 2015;56(8):4668–79.
12. Sauer L, et al. Monitoring macular pigment changes in macular holes using fluorescence lifetime imaging ophthalmoscopy. Acta Ophthalmol. 2017;95(5):481–92.
13. Dysli C, et al. Quantitative analysis of fluorescence lifetime measurements of the macula using the fluorescence lifetime imaging ophthalmoscope in healthy subjects. Invest Ophthalmol Vis Sci. 2014;55(4):2106–13.
14. Sharifzadeh M, Bernstein PS, Gellermann W. Nonmydriatic fluorescence-based quantitative imaging of human macular pigment distributions. J Opt Soc Am A Opt Image Sci Vis. 2006;23(10):2373–87.
15. Sauer L, et al. Fluorescence Lifetime Imaging Ophthalmoscopy (FLIO) of macular pigment. Invest Ophthalmol Vis Sci. 2018;59(7):3094–103.
16. Lakowicz JR. Principles of fluorescence spectroscopy. New York: Springer; 2007.
17. Sauer L, Gensure RH, Andersen K, Kreilkamp L, Hageman G, Hammer M, Bernstein PS. Patterns of fundus autofluorescence lifetimes in eyes of individuals with non-exudative age-related macular degeneration. Invest Ophthalmol Vis Sci. 2018;59:AMD65-AMD77.
18. Sauer L, Gensure RH, Hammer M, Bernstein PS. Fluorescence Lifetime Imaging Ophthalmoscopy (FLIO) – a novel way to assess Macular Telangiectasia Type 2 (MacTel). Ophthalmol Retina. 2017.
19. Dysli C, et al. Fluorescence lifetime imaging in stargardt disease: potential marker for disease progression. Invest Ophthalmol Vis Sci. 2016;57(3):832–41.
20. Gass JDM. Stereoscopic atlas of macular diseases: diagnosis and treatment. Missouri: Mosby; 1997.
21. Dithmar S. Macular hole. Survey and relevant surgical concepts. Ophthalmologe. 2005;102(2):191–206; quiz 207.
22. Kelly NE, Wendel RT. Vitreous surgery for idiopathic macular holes. Results of a pilot study. Arch Ophthalmol. 1991;109(5):654–9.
23. Kanski JJ. Klinische Ophthalmologie: Lehrbuch und Atlas. München: Urban & Fischer Verlag/ Elsevier Gmb; 2008.
24. Jordan F, et al. Study on the time course of macular pigment density measurement in patients with a macular hole--clinical course and impact of surgery. Klin Monatsbl Augenheilkd. 2012;229(11):1124–9.
25. Kishi S, Kamei Y, Shimizu K. Tractional elevation of Henle's fiber layer in idiopathic macular holes. Am J Ophthalmol. 1995;120(4):486–96.
26. Gass JD, Van Newkirk M. Xanthic scotoma and yellow foveolar shadow caused by a pseudo-operculum after vitreofoveal separation. Retina. 1992;12(3):242–4.
27. Ezra E, et al. Macular hole opercula. Ultrastructural features and clinicopathological correlation. Arch Ophthalmol. 1997;115(11):1381–7.
28. Harvey PS, King RA, Summers CG. Spectrum of foveal development in albinism detected with optical coherence tomography. J AAPOS. 2006;10(3):237–42.
29. Gregor Z. The perifoveal vasculature in albinism. Br J Ophthalmol. 1978;62(8):554–7.
30. Abadi RV, Cox MJ. The distribution of macular pigment in human albinos. Invest Ophthalmol Vis Sci. 1992;33(3):494–7.
31. Sparrow JR, Hicks D, Hamel CP. The retinal pigment epithelium in health and disease. Curr Mol Med. 2010;10(9):802–23.
32. Wolfson Y, et al. Evidence of macular pigment in the central macula in albinism. Exp Eye Res. 2016;145:468–71.
33. Putnam CM, Bland PJ. Macular pigment optical density spatial distribution measured in a subject with oculocutaneous albinism. J Optom. 2014;7(4):241–5.
34. Sauer L, Andersen KM, Gensure RH, Hammer M, Bernstein PS. Fluorescence Lifetime Imaging Ophthalmoscopy (FLIO) of macular pigment. Invest Ophthalmol Vis Sci. 2018;59:3094.

Chapter 17
Fluorescence Lifetime Imaging Ophthalmoscopy in Alzheimer's Disease

SriniVas R. Sadda

Alzheimer's disease (AD) is a chronic, progressive neurodegenerative disease which results in severe cognitive impairment. AD is the most frequent neurodegenerative disorder, and Alzheimer's type dementia is the common cause of dementia (60~70%). The risk of developing the disease increases every year with the growth of life expectancy. Globally, there are almost 46 million people living with dementia, and the number is expected to rise to 131.5 million by the year 2050 [1]. The definitive diagnosis of AD is only confirmed post mortem through an autopsy of the brain, by the histopathological identification of the hallmark proteolytic products of amyloid precursor protein (APP), β- amyloid (Aβ), and intracellular neurofibrillary tangles. In general, AD is a clinical diagnosis based on the criteria established by the National Institute of Neurological and Communicative Disorders and Stroke (NINCDS) and the Alzheimer's Disease and Related Disorders Association (ADRDA), and Diagnostic and Statistical Manual of Mental Disorders (DSM)-IV criteria [2]. It is well known that the pathophysiological process of AD begins years prior to the diagnosis of clinical dementia. Currently, significant efforts have been made in the development of diagnostic tools for detecting the presence and accumulation of Aβ biomarkers in the early stage of AD. Given that several clinical trials aimed at the stage of mild to moderate dementia have failed to demonstrate clinical benefit, the long preclinical phase of AD may prove to be a better target for disease-modifying therapy [3]. Recently, therapeutic interventions designed to slow down the aggregation of Aβ have been recommended to patients deemed to present with signs of preclinical AD, preceding mild cognitive impairment (MCI) or confirmed AD. Such interventions are targeted to help patients at a very early stage of the disease [4]. At present, Aβ and tau protein in cerebrospinal fluid (CSF), fluorodeoxyglucose (FDG)- and Pittsburg Compound B (PiB)-Positron Emission Tomography (PET) in the brain

S. R. Sadda (✉)
Doheny Eye Institute, University of California and Department of Ophthalmology, David Geffen School of Medicine at UCLA, Los Angeles, CA, USA
e-mail: ssadda@doheny.org

© Springer Nature Switzerland AG 2019
M. Zinkernagel, C. Dysli (eds.), *Fluorescence Lifetime Imaging Ophthalmoscopy*, https://doi.org/10.1007/978-3-030-22878-1_17

are thought to be useful for early diagnosis [5]. However, these methods are expensive, and/or invasive, and some of them require repeat exposure to radiation.

The retina is a central nervous system (CNS) tissue and shares many similar structural and functional features with the brain. Through the utilization of non-invasive imaging, the retina may be able to reflect the pathological changes in the brain. The first histological evidence of retinal abnormality in the human AD eye was widespread ganglion cell loss, retinal nerve fiber layer thinning, and optic nerve degeneration [6]. Since then, decreased retinal venous blood flow, retinal ganglion cell (RGC) degeneration in the peripheral retina, and Aβ deposition in retina have been reported [7–9]. Although Aβ plaque is the hallmark pathology of AD, Aβ deposition in retina is hard to visualize. While Koronyo-Hamaoui et al. [9–11] reported that early manifestations of retinal Aβ plaque pathology may be revealed in AD patients and animal models with a contrast agent such as curcumin, a few studies have failed to detect Aβ in the human AD retina [12, 13]. However, there is growing evidence that the presence of Aβ plaque in retinas of AD patients is comparable with that observed in their brains, and it has also been shown that the presence of Aβ plaque in patients with AD is correlated with retinal structural deficits such as retinal ganglion cell loss, nerve fiber layer atrophy, and thinning of the macular ganglion cell complex [6–9].

Fluorescence lifetime measurements offer a potentially rich dataset with the ability to resolve many different retinal fluorophores which may allow for the detection of retinal abnormalities at an early stage. Therefore, the detection of subtle changes and discrimination of these fluorophores might be helpful in detecting subtle alterations in the retina in Alzheimer's patients. Most FLIO studies at present, utilize a modified Spectralis system (Heidelberg engineering) which features a blue pulsed laser (473 nm; 80 MHz; 70 ps pulse width) and highly sensitive hybrid detectors which allow detection in two separate channels: a short spectral channel (SSC; 498–560 nm) and a long spectral channel (LSC: 560–720 nm). Precise single photon counting cards keep track and quantify the number and time of detection of photons.

In a pilot study, Jentsch et al. reported that FLIO measurements may correlate with some Alzheimer-specific markers [14]. Specifically, they noted that the mini-mental state examination (MMSE) score and CSF tau protein showed a significant correlation with $\alpha 2$ and Q2 in the LSC FLIO parameters [14].

In another FLIO study by Kwon et al., focusing primarily on subjects with an early stage of Alzheimer's patients, Alzheimer-associated laboratory data, OCT data and FLIO-derived parameters were compared and correlated between the preclinical Alzheimer's group and controls [15]. τ_m of the macular retina in the AD group showed significant differences compared to control group (longer lifetimes in AD subjects), and correlated with Aβ, tau level in the CSF, and the ganglion cell layer plus inner plexiform layer (GCL/IPL) thickness (Fig. 17.1). Of note, as in the previous study, the correlations of the clinical and OCT data with FLIO were stronger with FLIO parameters in the LSC than in the SSC. This is perhaps not surprising as it is well-established that SSC parameters are much more confounded by the lens status, as the crystalline lens (esp. a cataractous lens) tends to absorb short-wavelength light. In contrast, the LSC is relatively unaffected by the lens status [16]. As patients with preclinical AD are generally older and frequently have media opacity, lifetime

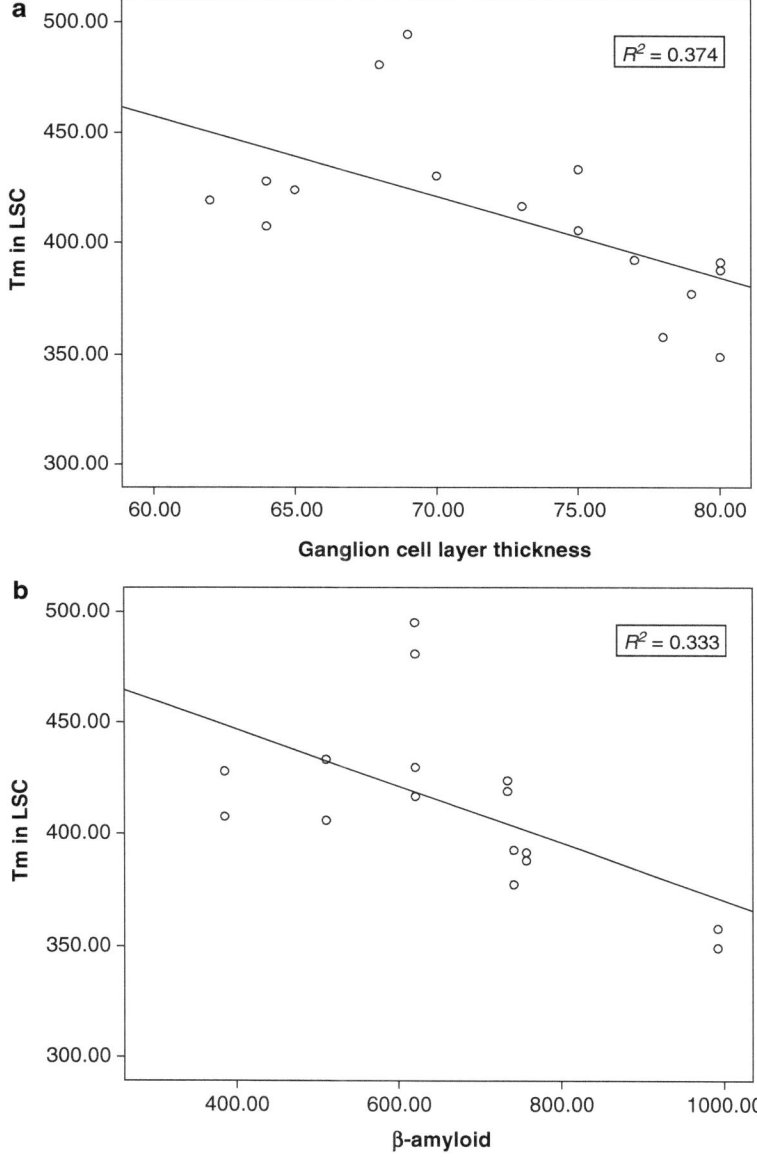

Fig. 17.1 Figure Correlation plot of mean fluorescence lifetime (τ_m) with clinical data in Alzheimer's patients. (**a**) Correlation plot of τ_m of LSC with GCL + IPL thickness. (**b**) Correlation plot of τ_m of LSC with amyloid β level in CSF. (**c**) Correlation plot of τ_m of LSC with total tau level in CSF

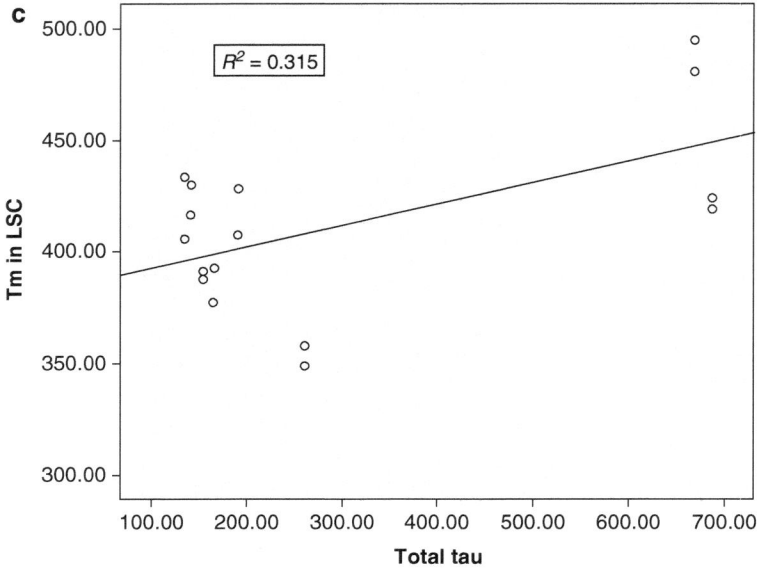

Fig. 17.1 (continued)

data from the LSC may be of most value for studying Alzheimer' subjects. The mechanism for the prolonged lifetimes in early/ pre-clinical AD subjects is at present unknown. One may speculate that this may correlate with tau protein or Aβ amyloid in the retina, but this would need to be confirmed with histologic analysis.

Of note, Kwon's study did not observe a correlation between fluorescence lifetime and vascular density of the retinal capillary plexuses or the choriocapillaris on OCT angiography (OCTA), suggesting that FLIO measurements may provide additional independent information not available in the OCTA data. It should be noted, however, that these OCTA results are in contrast to Bulut et al. who reported that the foveal avascular zone was enlarged and retinal vascular density was reduced on OCTA in AD patients [17]. They also observed that the choroidal thickness was lower in AD patients compared with normal controls. Regardless, correlation of fluorescence lifetime with vascular density in AD, or preclinical AD, needs further evaluation with larger cohorts.

Summary Box

In summary, early pilot studies suggest that FLIO abnormalities, primarily prolonged lifetimes in the long spectral channel, may be observed in pre-clinical/early Alzheimer's Disease spectrum patients. The molecular basis and pathophysiology underlying these changes has yet to be established. Nonetheless, if these preliminary observations can be replicated and validated in future larger, prospective, longitudinal studies, FLIO may prove to be useful as a non-invasive diagnostic tool for Alzheimer's disease.

References

1. Prince PM, Ali G, Ali G. World Alzheimer report 2015. The global impact of dementia. London: Alzheimers Disease International; 2015.
2. McKhann GM, Knopman DS, Chertkow H, Hyman BT, Jack CR Jr, Kawas CH, et al. The diagnosis of dementia due to Alzheimer's disease: recommendations from the National Institute on Aging-Alzheimer's Association workgroups on diagnostic guidelines for Alzheimer's disease. Alzheimers Dement. 2011;7:263–9. https://doi.org/10.1016/j.jalz.2011.03.005.
3. Sperling RA, Aisen PS, Beckett LA, Bennett DA, Craft S, Fagan AM, et al. Toward defining the preclinical stages of Alzheimer's disease: recommendations from the National Institute on Aging-Alzheimer's Association workgroups on diagnostic guidelines for Alzheimer's disease. Alzheimers Dement. 2011;7:280–92. https://doi.org/10.1016/j.jalz.2011.03.003.
4. Lim YY, Maruff P, Getter C, Snyder PJ. Discloure of PET amyloid imaging results: a preliminary study of safety and tolerability. Alzheimers Dement. 2016;12:454–8. https://doi.org/10.1016/j.jalz.2015.09.005.
5. Rabinovici GD, Rosen HJ, Alkalay A, Kornak J, Furst AJ, Agarwal N, et al. Amyloid vs FDG-PET in the differential diagnosis of AD and FTLD. Neurology. 2011;77:2034–42. https://doi.org/10.1212/WNL.0b013e31823b9c5e.
6. Hinton DR, Sadun AA, Blanks JC, Miller CA. Optic nerve degeneration in Alzheimer's disease. N Engl J Med. 1986;315:485–7.
7. Feke GT, Hyman BT, Stern RA, Pasquale LR. Retinal blood flow in mild cognitive impairment and Alzheimer's disease. Alzheimers Dement. 2015;1:144–51. https://doi.org/10.1016/j.dadm.2015.01.004.
8. Bambo MP, Garcia-Martin E, Pinilla J, Herrero R, Satue M, Otin S, et al. Detection of retinal nerve fiber layer degeneration in patients with Alzheimer's disease using optical coherence tomography:searching new biomarkers. Acta Ophthalmol. 2014;92:581–2. https://doi.org/10.1111/aos.12374.
9. Koronyo-Hamaoui M, Koronyo Y, Ljubimov AV, Miller CA, Ko MK, Black KL, et al. Identification of amyloid plaques i retinas from Alzheimer's patients and noninvasive in vivo optical imaging of retinal plaques in a mouse model. NeuroImage. 2011;54(Suppl 1):S204–17. https://doi.org/10.1016/j.neuroimage.2010.06.020.
10. Koronyo Y, Salumbides BC, Black KL, Koronyo-Hamaoui M. Alzheimer's disease in the retina:imaging retinal A plaques for early diagnosis and therapy assement. Neurodegener Dis. 2012;10:285–93. https://doi.org/10.1159/000335154.
11. Koronyo Y, Biggs D, Barron E, Boyer DS, Pearlman JA, Au WJ, Kile SJ, et al. Retinal amyloid pathology and proof-of concept imaging trial in Alzheimer's disease. JCI Insight. 2017;17:93621. https://doi.org/10.1172/jci.insight.93621.
12. Schön C, Hoffmann NA, Ochs SM, Burgold S, Filser S, Steinbach S, et al. Long-term in vivo imaging of fibrillar tau in the retina of P301S transgenic mice. PLoS One. 2012;7:e53547. https://doi.org/10.1371/journal.pone.0053547.
13. Ho CY, Troncoso JC, Knox D, Stark W, Eberhart CG. Beta-amyloid, phospho-tau and alpha-synuclein deposits similar to those in the brain are not identified in the eyes of Alzheimer's and Parkinson's disease patients. Brain Pathol. 2014;24:25–32. https://doi.org/10.1111/bpa.12070.
14. Jentsch S, Schweitzer D, Schmidtke KU, Peters S, Dawczynski J, Bär KJ, et al. Retinal fluorescence lifetime imaging ophthalmoscopy measures depend on the severity of Alzheimer's disease. Acta Ophthalmol. 2015;93:e241–7. https://doi.org/10.1111/aos.12609.
15. Kwon S, Fan W, Borreli E, et al. Fluorescence lifetime imaging ophthalmoscopy in early Alzheimer's disease patients. 2018 ARVO annual meeting.
16. Dysli C, Wolf S, Zinkernagel MS. Autofluorescence lifetimes in geographic atrophy in patients with age-related macular degeneration. Invest Ophthalmol Vis Sci. 2016;57:2479–87. https://doi.org/10.1167/iovs.15-18381.
17. Bulut M, Kurtuluş F, Gözkaya O, Erol MK, Cengiz A, Akıdan M, et al. Evaluation of optical coherence tomography angiographic findings in Alzheimer's type dementia. Br J Ophthalmol. 102:233. https://doi.org/10.1136/bjophthalmol-2017-310476.

Chapter 18
FLIO in Mouse Models

Chantal Dysli, Muriel Dysli, and Martin Zinkernagel

Fluorescence lifetime imaging ophthalmoscopy has been proven to be a reliable technique to investigate fluorescence lifetimes in human retina in healthy eyes and in retinal diseases [1, 2]. However, measured fluorescence lifetimes have to be interpreted in the complex context of the living eye. Thereby, the variety of contribution of different structures of the eye, especially individual retinal layers turns out to be challenging. Furthermore, contribution of various cell types, molecules, molecular interaction, and pathways is expected. In order to bridge the gap between in vitro measurements using FLIO or FLIM to the clinical setting, the FLIO technique has been established and optimized for the measurement in animal models such as mice by Dysli et al. [3, 4] and was later confirmed in rats [5].

Technically, the same FLIO setup was used as for measurements in human eyes. Additionally, the camera was adjusted by adding a 25-diopter (D) lens (f = 40/+25 D; Heidelberg Engineering, Heidelberg, Germany) in front of the FLIO aperture in order to adapt the short axial length of the mouse eye (e.g. 3 mm).

FLIO imaging in the murine fundus revealed a characteristic distribution of fluorescence lifetimes. In contrast to the human retina, in mice, the shortest fluorescence lifetimes were measured in the area of retinal vessels (SSC: 600–800 ps [mouse] vs 415 ps [human]; LSC: 250–350 ps vs 350 ps). The surrounding retina featured mean fluorescence lifetimes of about 950 ps (for SSC; LSC = 300 ps) in pigmented mice (C57BL/6), and 800 ps (LSC = 250 ps) in non-pigmented albino mice (BALBc) (Fig. 18.1). In comparison to the human retina, measured lifetime values in mice were much higher. Cohorts of healthy animals were imaged monthly with FLIO and OCT with a total followed-up of 6 months. Finally, comparison with histology was performed. Over time and advanced age of the mice, a decrease of fluorescence lifetimes was measured in pigmented as well as in non-pigmented animals. This is in contrast

C. Dysli (✉) · M. Dysli · M. Zinkernagel
Department of Ophthalmology and Department of Clinical Research, Inselspital,
Bern University Hospital, University of Bern, Bern, Switzerland
e-mail: chantal.dysli@insel.ch

© Springer Nature Switzerland AG 2019
M. Zinkernagel, C. Dysli (eds.), *Fluorescence Lifetime Imaging Ophthalmoscopy*, https://doi.org/10.1007/978-3-030-22878-1_18

114

C. Dysli et al.

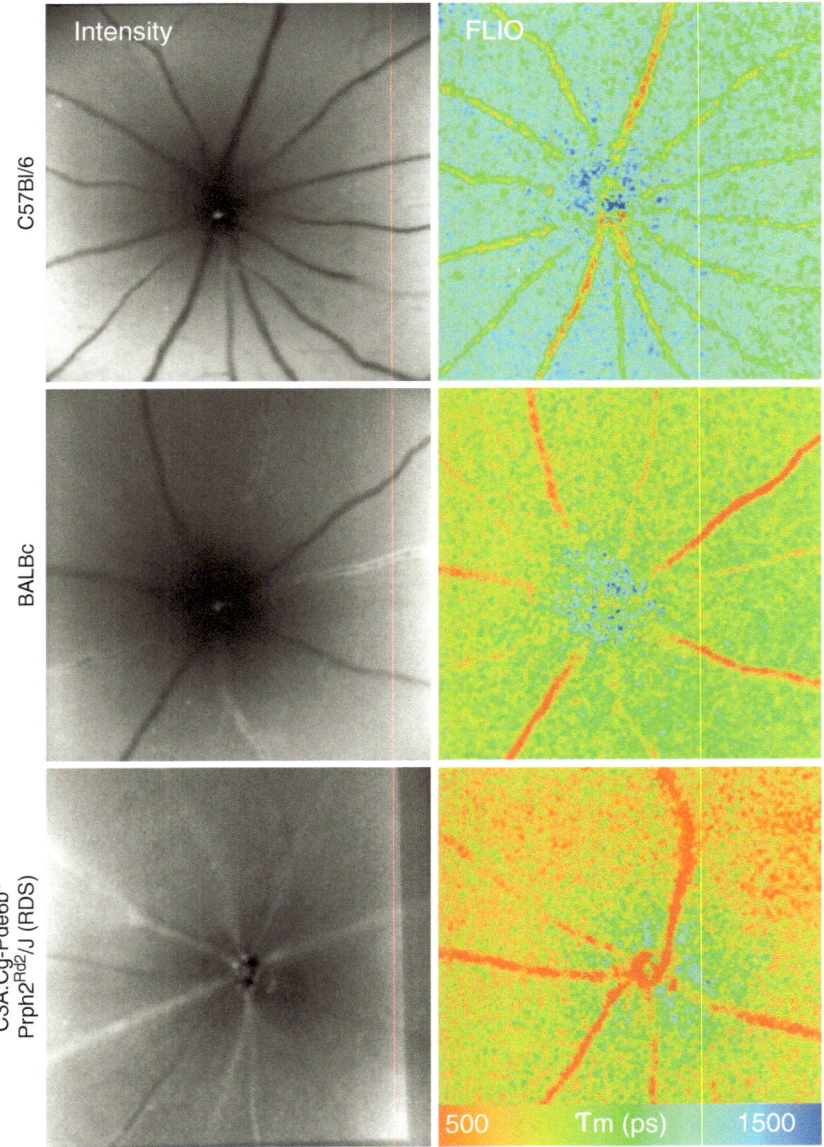

Fig. 18.1 FLIO imaging in three mouse model: pigmented C57Bl/6 mouse, nonpigmented BALBc mouse, and C3A.Cg-Pde6b⁵Prph2^Rd2/J (retinal degeneration slow, RDS) mouse. Fluorescence lifetimes within the murine retina is much longer compared to the human retina. (note: adjusted color scale). However, retinal vessels appeared with shorter lifetimes compared to the surrounding retina

to the steadily increase of fluorescence lifetimes in the human retina with age. However, fluorescence lifetimes of aged mice and human retina finally reach similar mean lifetime values. Possibly, this reflects a composition of similar fluorophores within both, the human and the murine retina. Regarding the pigmentation of the fundus, albino mice generally featured shorter mean fluorescence lifetime compared to pigmented mice. Therefore, a correlation to the amount of melanin, which showed relatively long fluorescence lifetimes when measured in vitro, can be assumed.

Additionally, Dysli et al. investigated a mouse model of slow retinal degeneration (RDS, C3A.Cg-Pde6b^{5}Prph2^{Rd2}/J). Over time, progression of patchy retinal changes was observed, leading to prolongation of retinal fluorescence lifetimes. In one moment in time of measurement, a wide range of lifetime values was observed within one eye as well as between different animals, mirroring the different stages of retinal degeneration.

In order to differentiate the contribution of different retinal layers to the measured mean fluorescence lifetime, Dysli et al. pharmacologically induced retinal degeneration [2]. Degeneration of the RPE followed by subsequent loss of photoreceptors was induced by sodium iodate (NaIO$_3$) [6]. Specific degeneration of photoreceptors with preservation of the RPE was induced using N-methyl-N-nitrosourea (MNU) [7]. All animals were measured weekly with FLIO and OCT over a time period of 4 weeks. In case of RPE- and subsequent photoreceptor degeneration induced by NaIO$_3$, a prolongation of the mean fluorescence lifetime was observed. On the other hand, if the RPE was preserved and only the photoreceptors were targeted and destroyed by NMU, shorter fluorescence lifetime values were measured.

Dysli et al. concluded that short fluorescence lifetimes in the murine retina might origin from the RPE and might be influenced and modulated by the overlying neurosensory retina. However, in absence of the RPE and the photoreceptors, increased contribution of other retinal layers- such as the choroid- might be detected which feature relatively long fluorescence lifetimes.

In general, FLIO measurements in mouse or rat models was shown to be a reproducible tool to investigate retinal fluorescence lifetimes beyond the human retina. The main advantages are that specific structural changes and metabolic conditions can be induced and targeted. Additionally, well characterized hereditary conditions can be easily investigated using FLIO. Especially follow-up examinations over time were shown to be of great interest to investigate the sequence of retinal changes in FLIO chronologically. Thereby, identification of individual fluorophores and metabolic interactions and basic pathophysiological pathways could be simplified. However, results have to be interpreted with caution as they might not be directly comparable and transferable to FLIO measurements in the human eye. Nevertheless,

FLIO in mouse models might also proof useful as a monitoring tool in therapeutic trials to investigate treatment effects and their outcome over time.

Summary Box
FLIO in animal models provides a promising and well-controlled method for further investigation of physiological, structural, pharmacological, and genetical changes within the retina, and for follow-up over time, possibly also in terms of future therapeutical approaches.

References

1. Dysli C, et al. Quantitative analysis of fluorescence lifetime measurements of the macula using the fluorescence lifetime imaging ophthalmoscope in healthy subjects. Invest Ophthalmol Vis Sci. 2014;55(4):2106–13.
2. Dysli C, et al. Fluorescence lifetime imaging ophthalmoscopy. Prog Retin Eye Res. 2017;60:120–43.
3. Dysli C, et al. Fluorescence lifetime imaging of the ocular fundus in mice. Invest Ophthalmol Vis Sci. 2014;55(11):7206–15.
4. Dysli C, et al. Effect of pharmacologically induced retinal degeneration on retinal autofluorescence lifetimes in mice. Exp Eye Res. 2016;153:178–85.
5. Teister J, et al. Peripapillary fluorescence lifetime reveals age-dependent changes using fluorescence lifetime imaging ophthalmoscopy in rats. Exp Eye Res. 2018;176:110–20.
6. Enzmann V, et al. Behavioral and anatomical abnormalities in a sodium iodate-induced model of retinal pigment epithelium degeneration. Exp Eye Res. 2006;82(3):441–8.
7. Tsubura A, et al. Animal models for retinitis pigmentosa induced by MNU; disease progression, mechanisms and therapeutic trials. Histol Histopathol. 2010;25(7):933–44.

Chapter 19
Discussion and Summary

Clinical Applications, Limitations, Outlook, Future Perspectives

Chantal Dysli, Martin Zinkernagel, and Lydia Sauer

FLIO adds a number of benefits to traditional fundus autofluorescence imaging. There is emerging evidence that with FLIO hidden details of the visual cycle can be exposed. Thus, FLIO potentially provides information about the viability of photoreceptors in retinal diseases. In the last years, reports on fluorescence lifetime imaging of macular pigment, retinal dystrophies, and several degenerative diseases have been published and have brought new insights into the use of FLIO for retinal diagnostics. Important steps have been made to establish acquisition protocols and standards for data analysis to compare results between centers.

However, currently there are still several limitations associated with the FLIO technology. Primarily, data analysis is complex and time consuming. This makes it challenging to use the device in clinical routine. We hope that this issue can be resolved by automated algorithms in the future. Other limitations include lifetime data shifts caused by opacities of the optical system within the eye, especially cataracts of the crystalline lens. FLIO imaging is still possible in patients with cataracts, but images need to be analyzed more critically and may not be comparable to age-matched healthy eyes without cataracts. In addition, the paucity of devices available has so far hindered the widespread use of FLIO for research purposes or clinical use. With a more widespread availability of FLIO cameras, this limitation will hopefully be resolved in the near future. With refinements made to the currently available system, better resolution of lifetime data can be expected. A useful addition would be adjustable excitation wavelength and emission spectra in order to obtain spectral information about individual fluorophores. With this, the detection

C. Dysli · M. Zinkernagel (✉)
Department of Ophthalmology and Department of Clinical Research, Inselspital, Bern University Hospital, University of Bern, Bern, Switzerland
e-mail: martin.zinkernagel@insel.ch

L. Sauer
University of Utah, John A. Moran Eye Center, Salt Lake City, UT, USA

© Springer Nature Switzerland AG 2019
M. Zinkernagel, C. Dysli (eds.), *Fluorescence Lifetime Imaging Ophthalmoscopy*, https://doi.org/10.1007/978-3-030-22878-1_19

method would need refinements as well, in order to obtain detailed spectral information about individual photons before rendering this information into lifetime data. Combining FLIO with quantitative FAF (qAF) would allow obtaining more detailed information about lipofuscin levels in the RPE.

Nevertheless, when looking back at the last decade of FLIO research, a tremendous amount of new knowledge has been gained. FLIO allows for detection of all retinal diseases that can be identified with conventional imaging technologies. However, FLIO takes retinal imaging a step further. Beyond detecting the nutritional status of the eye by imaging macular pigment, FLIO holds additional information about disease processes. This has been shown highly valuable for a large number of diseases. In Stargardt disease, onset and progression of new flecks can be detected even before any corresponding signs are visible in other imaging technologies. In MacTel, the diagnosis of the disease can be made with FLIO before other imaging techniques can detect any abnormalities. Furthermore, FLIO shows higher contrast than other imaging modalities. In age related macular degeneration, specific patterns of FLIO lifetimes may indicate the presence or risk to develop a more advanced disease stage. Many other diseases have yet to be investigated to show the full potential of FLIO in the clinical realm. With further refinements in this technique, it may take an even larger part in the currently available retinal imaging armamentarium. Overall, FLIO emerges as a great addition to existing conventional retinal imaging techniques, and may find an application in clinical routine in the future.

Index

© Springer Nature Switzerland AG 2019
M. Zinkernagel, C. Dysli (eds.), *Fluorescence Lifetime Imaging
Ophthalmoscopy*, https://doi.org/10.1007/978-3-030-22878-1